Chapter 1

Galaxies: dynamics, potential theory, and equilibria

1.1 Dynamics of scattering

A natural way to begin is to ask the seemingly naive question "what is a galaxy?" The straightforward answer is that a galaxy is a system of stars and gas, like the Milky Way (the word *galaxy* comes from the Greek for "milk"). Prior to the last century, the Milky Way was not just the only known galaxy, but was thought to constitute the entire Universe. In 1926, Edwin Hubble showed that the Andromeda nebula was actually a separate galaxy, spanning 3° in the sky, another blow to our anthropocentric view of the Universe (if one was needed). The Earth is not the center of the Solar System, the Solar System is not the center of the Milky Way, and the Milky Way is just one of billions (and billions) of galaxies in the Universe.

Just as stars are held up by internal gas pressure, galaxies also have internal energy that prevents gravitational collapse, held in equilibrium by the virial theorem (e.g. Hansen and Kawaler 2004),

$$2T + U = 0, \qquad (1.1)$$

where T is the internal kinetic energy of the particles (stars) and U is the negative gravitational potential energy. Like any self-gravitating system, the potential energy, up to constants of order unity, is given by

$$U \sim -\frac{GM^2}{R}, \qquad (1.2)$$

where M and R are the mass and characteristic size of the system and G is Newton's gravitational constant[1]. The kinetic energy is given by

$$T = \frac{1}{2}Mv^2, \qquad (1.3)$$

---

[1] Although G is known to relatively poor accuracy in cgs units, in combination with the solar mass $GM_\odot$ is known quite well. We can specify G in solar mass units: $G = 4.30091851 \times 10^{-3} M_\odot^{-1}(km/s)^2 pc$. This is *very* useful given astronomers' habit of quoting masses in terms of $M_\odot$ and distances in terms of parsecs.

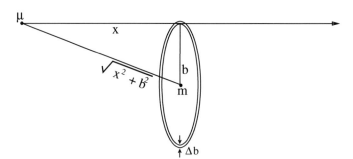

Figure 1.1: A star of mass μ travelling with velocity v passes a second star, with mass m at a distance of closest approach (impact parameter) b.

where v^2 is the average squared velocity for a particle in the system. The *gravitational potential* is then of order

$$\Phi \sim \frac{GM}{R} \sim v^2. \tag{1.4}$$

For a typical galaxy we have:

radius $R \sim 10$ kpc

density $\rho \sim 0.1 M_\odot/\text{pc}^3$ and,

$N_{\text{stars}} \sim 10^{11}$.

The "thermal" velocity for individual stars within the galaxy is then $v \sim 210$ km/s.

Unlike the interiors of stars, where the mean free path (mfp) of particles is much shorter than the size of the star (mfp $\sim 1\mu m \ll R$), the mean free path of a star within a galaxy is much *larger* than the size of the galaxy, suggesting that the average star cannot be expected to scatter off another star during an orbit or even during its lifetime. How do we quantify scattering interactions? For a star with mass μ moving along a trajectory $x = vt$ with impact parameter b (the distance of closest approach if there were no deflection; see Fig. 1.1) relative to another star with mass m, the change in the component of velocity perpendicular to the direction of travel, $\Delta v_\perp$ is given by

$$\Delta v_\perp = \frac{F_\perp \Delta t}{\mu} = \frac{Gmb}{(x^2+b^2)^{3/2}}\Delta t. \tag{1.5}$$

Assuming a constant speed v during the interaction, this change in perpendicular velocity can be integrated over time, giving

$$\Delta v_\perp = \int_{-\infty}^{\infty} \frac{F_\perp}{\mu} dt = \frac{Gm}{b^2}\int_{-\infty}^{\infty} \frac{dt}{\left[1+\left(\frac{vt}{b}\right)^2\right]^{3/2}} = 2\frac{Gm}{bv}. \tag{1.6}$$

We call a scattering "strong" if $\Delta v_\perp = v$, so $b_{\text{strong}} = 2Gm/v^2$. The strong scattering cross section σ is therefore

$$\sigma = \pi b_{\text{strong}}^2 = \pi \left(\frac{2Gm}{v^2}\right)^2 \sim \frac{4\pi}{N^2}R^2, \tag{1.7}$$

1.1. DYNAMICS OF SCATTERING

where we have used the virial theorem to get $v^2 \sim GmN/R$. We now want to see how often a star might actually experience a strong encounter with another star. For low number densities n (an excellent assumption for galaxies), the probability of a strong scattering encounter during a single orbital crossing is given by

$$\begin{pmatrix} \text{probability of a} \\ \text{strong encounter} \\ \text{in one crossing} \end{pmatrix} \approx n\sigma R = \frac{N}{\frac{4}{3}\pi R^3} \cdot \frac{4\pi}{N^2} R^2 \cdot R = \frac{3}{N}, \quad (1.8)$$

a very small number for galaxies with $N \sim 10^{11}$! So we can safely say that strong scattering does not play a very important role in the orbits of stars within the galaxy.

We must nonetheless consider the cumulative effects of weak scattering from long-range gravitational interactions. Treating the encounters as a collection of steps in a random walk, we find

$$(\Delta v_\perp)_{\text{total}} = \sqrt{\sum_i (\Delta v_{\perp i})^2} \quad (1.9)$$

Integrating over all possible impact parameters, we find that over a time Δt, the accumulated perpendicular velocity is given by

$$\sum_i (\Delta v_{\perp i})^2 = \int_0^\infty \underbrace{2\pi v \Delta t\, b\, db}_{\text{volume element}}\, n \left(\frac{2Gm}{bv}\right)^2 = \Delta t \frac{8\pi G m^2 n}{v} \int_0^\infty \frac{db}{b}. \quad (1.10)$$

In order to get a finite value for the integral, we must impose physical bounds on b. The upper limit is the size of the galaxy. The lower limit is the impact parameter at which a collision is judged to be strong. Taking $b_{\min} = Gm/v^2$ and $b_{\max} = R$, the integral can be evaluated as

$$\int_{b_{\min}}^{b_{\max}} \frac{db}{b} = \ln \Lambda = \ln \frac{Rv^2}{Gm} \sim \ln N \sim 25, \quad (1.11)$$

where we again use the virial theorem to substitute $R/Gm \sim N/v^2$.

For a system of stars in a galaxy (or globular cluster or any collisionless, gravitationally bound system), we can define a *relaxation time* for which the accumulated perpendicular velocity is comparable to the average velocity: $(\Delta v_\perp)_{\text{tot}} \sim v$. Combining equations (1.9, 1.10, and 1.11), we see that

$$t_{\text{relax}} \sim \frac{v^3}{8\pi G^2 n m^2} \frac{1}{\ln \Lambda}. \quad (1.12)$$

The number of orbits needed for the system to "relax" is given by

$$\frac{t_{\text{relax}}}{t_{\text{orbit}}} \sim \frac{v^3}{8\pi G^2 n m^2 \ln \Lambda} \frac{1}{R} \sim \frac{v^4 R^2 \frac{4}{3}\pi}{8\pi G^2 N m^2 \ln \Lambda} \sim \frac{N}{6 \ln N}, \quad (1.13)$$

so it seems that weak scattering is *also* relatively insignificant for calculating orbits within galaxies. Evidently we must think of a galaxy not as a large collection of scatterers, but rather as a gravitational field that is created by and interacts with the matter in the galaxy.

1.2 Building galaxies

As we saw above, gravitational scattering plays a very small role in the dynamics of stellar orbits within galaxies or globular clusters. We return to our original question, "what is a galaxy?" In light of the previous section, it seems better to think of a galaxy as a collisionless fluid of stars, a superposition of a large number of orbits guided by a single background potential.

How does one put together such a galaxy? Superpose orbits! Schwarzschild's method (see Figure 1.2) follows these simple steps:

- choose gravitational potential
- generate an orbit library
- populate potential with orbits
- check for self-consistency

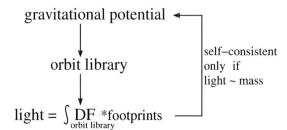

Figure 1.2: Schwarzschild's method for building galaxies. One adjusts the distribution function (DF) to match the light distribution or the gravitational potential.

For example, imagine building a spherically symmetric galaxy. How would the orbits be distributed in phase space? Even though they may occupy three-dimensional space quite uniformly there is no reason to insist that the orbits should be uniformly distributed in momentum space. A galaxy made only of circular orbits would appear "hot" in the tangential direction and "cold" in the radial direction, while a collection of radial orbits (an equally valid approach to a spherical system) would look hot in the radial direction and cold in the tangential direction (see Figure 1.3).

 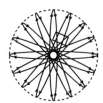

Figure 1.3: Two schemes for constructing a spherical galaxy: out of circular orbits, left, and out of (nearly) radial orbits, right.

"Hot" and "cold" can be understood quantitatively by defining a symmetric *velocity dispersion* tensor:

$$\sigma_{ij}^2 = \overline{(v_i - \overline{v}_i)(v_j - \overline{v}_j)}, \tag{1.14}$$

where $\overline{v}_i$ is the velocity in the $\hat{\mathbf{e}}_i$ direction averaged at some point. For a collisionless fluid, we can diagonalize the velocity dispersion tensor, giving three independent temperatures at each point in

1.2. BUILDING GALAXIES

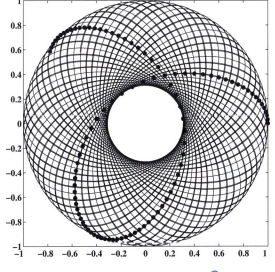

Figure 1.4: The orbit of a star on a spherically symmetric logarithmic potential. The position of the star is plotted for the first 100 time steps. By following the star for many orbits we accumulate a probability density "footprint" that may be used to generate a galaxy model.

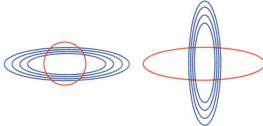

Figure 1.5: Left: a disk in circular orbit around a spherical bulge. Right: a disk in polar orbit around a triaxial bulge. Such "polar ring" galaxies constitute perhaps 0.1% of all galaxies.

space. If the random velocities are greater than (or in the same order as) the ordered velocities, the fluid is *hot*; if the ordered velocities are much greater, the fluid is *cold*.

Observations (Figure 1.7) show that many galaxies have nearly constant rotational velocities with $v(r) \approx v_c$ over the observable range of radius. These flat rotation curves are a departure from Keplerian orbital velocities, which would scale as $v(r) \sim r^{-1/2}$. Flat rotation curves imply mass increasing linearly with radius, and so cannot continue without bound.

For a spherically symmetric system, the centripetal acceleration is given by

$$\frac{v^2}{r} = \frac{d\Phi}{dr} \tag{1.15}$$

For a constant rotation velocity $v(r) = v_c$, the solution for the potential is

$$\Phi = v_c^2 \ln \frac{r}{r_{\text{ref}}}. \tag{1.16}$$

Combining with Poisson's equation $\nabla^2 \Phi = 4\pi G\rho$, we find an expression for the density profile:

$$\rho(r) = \frac{v_c^2}{4\pi G r^2}. \tag{1.17}$$

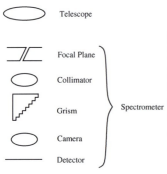

Figure 1.6: A schematic representation of a spectrograph. Light from the telescope is brought to a focus in the focal plane. The slit of the spectrograph isolates the light from a narrow strip of the sky, which diverges and passes through the collimator. Parallel light rays from the collimator enter the grism (a grating ruled on a prism) which disperses the different wavelengths. The camera brings the light to a focus on the detector.

Figure 1.7: Left: The slit of a spectrometer is placed along the major axis of a disk galaxy. The disk is highly inclined to the line of sight. Stars and gas on the left side of the slit have a line of sight velocity toward us (relative to the center of the galaxy). Right: The spectrum of the galaxy. The lines shown are typical of gas photo-ionized by hot stars. The lines are blueshifted on the left side and redshifted on the right side. This gives the *rotation curve* of the galaxy. The rotation curve is flat in the outer parts of the galaxy and roughly linear in the inner parts.

1.2. BUILDING GALAXIES

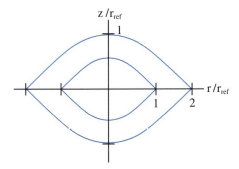

Figure 1.8: Equipotentials for Mestel's Disk

Among the obvious shortcomings of this model are the failure of the potential to go to zero at large radius and the prediction of infinite mass. Notwithstanding, it is still a useful potential for many practical applications.

Another weakness of this model is that it is limited to spherically symmetric galaxies and we know from experience that many galaxies are shaped more like flattened disks. A corresponding model for such a galaxy is called a *Mestel disk*, which is azimuthally symmetric around the z-axis. The potential for the Mestel disk is

$$\Phi(r,\theta) = v_c^2 \left[\ln \frac{r}{r_{\text{ref}}} + \ln \frac{1+|\cos\theta|}{2} \right], \qquad (1.18)$$

where θ is measured from the positive z-axis. The Laplacian in spherical geometry with azimuthal symmetry is given by

$$\nabla^2 = \frac{1}{r^2}\frac{\partial}{\partial r}r^2\frac{\partial}{\partial r} + \frac{1}{r^2 \sin\theta}\frac{\partial}{\partial \theta}\sin\theta\frac{\partial}{\partial \theta}. \qquad (1.19)$$

Applying Poisson's equation to equation (1.18), we find that the Mestel disk has zero density everywhere except $\theta = \pi/2$, where the density is infinite. The more useful (and in this case finite) quantity is the *surface density*, determined by integrating the density over the (infinitesimal) thickness of the disk:

$$\Sigma(r) = \int \rho(r,z) dz = \frac{v_c^2}{2\pi G r}. \qquad (1.20)$$

Like the spherical model for a flat rotation curve, the disk model also suffers from an infinite total mass and radius, but it is still useful for many applications.

Observations show that most galaxies have elements of *both* of the above models, with a roughly spherical bulge *and* a flattened disk. The relative weight of these features (the "disk-to-bulge ratio," Figure 1.9) is a useful parameter for classifying galaxies, with disk strength correlating positively with other galaxy properties such as young stellar fraction, hydrogen fraction, CO gas, dust, and the strength of spiral arms. In most galaxies the spiral arms, which are located within the disk are the primary location for new star formation. The disk and bulge components of most galaxies appear to have had rather different histories, a subject to be raised again in the context of galaxy formation.

Soon after the realization that many nebulae were in fact distinct galaxies, astronomers (and in particular Hubble) began systematically classifying galaxies in the hope that taxonomy would lead to further understanding. The principal component of Hubble's classification scheme is strongly correlated with disk-to-bulge ratio. But Hubble conflated the relative proportions of bulge and disk

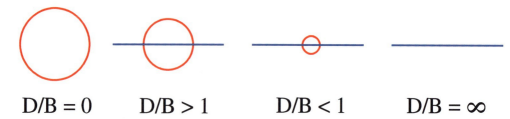

D/B = 0   D/B > 1   D/B < 1   D/B = ∞

Figure 1.9: Galaxies with varying disk to bulge ratios, seen edge on.

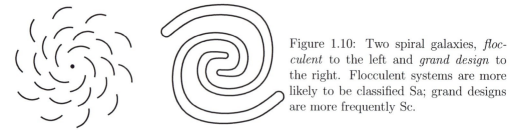

Figure 1.10: Two spiral galaxies, *flocculent* to the left and *grand design* to the right. Flocculent systems are more likely to be classified Sa; grand designs are more frequently Sc.

with the prominence of spiral arms within the disk. In Hubble's scheme most galaxies were classified as "E", "Sa", "Sb" or "Sc." "E" stood for elliptical with no sign of spiral structure. "S" stood for spiral structure, with "Sa" galaxies showing the least spiral structure and "Sc" galaxies showing the most, as shown in Figure 1.10. The sequence from E to Sc is approximately one of increasing disk-to-bulge ratio. But there are also galaxies with prominent disks and little evidence for spiral structure. Hubble added the S0 classification, placing it between E and Sa. Modern versions of Hubble's scheme have put the S0 systems parallel to the S systems, with S0/a, S0/b and S0/c systems having increasingly large disk-to-bulge ratios. One might hope that disk-to-bulge ratio would completely replace Hubble's system, but while easy to describe, disk-to-bulge ratio can be difficult to measure, particularly for systems that are face-on. Modern "quantitative" classification schemes often use central concentration as a proxy for disk-to-bulge ratio and use deviation from bilateral symmetry as a proxy for spiral structure.

1.3 Extragalactic empiricism

Before we can discuss galaxies quantitatively, we must first introduce some new terms. The *specific intensity* $I_\nu(\vec{\theta})$ is a measure of energy flux incident on a unit surface area per unit solid angle and frequency (erg/s/cm^2/sterad/Hz). Then the *specific flux* on a given surface element is the specific intensity integrated over all solid angle:

$$f_\nu = \int I_\nu d\Omega. \tag{1.21}$$

For the purpose of the present discussion, the units of specific flux are ergs/s/cm^2/Hz. The *specific luminosity* L_ν at a given frequency ν of an object at a distance D is then

$$L_\nu = 4\pi D^2 f_\nu. \tag{1.22}$$

(Continued on next page.)

1.3. EXTRAGALACTIC EMPIRICISM

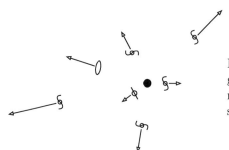

Figure 1.11: The observed velocities of nearby galaxies are proportional to their distances, permitting distances to be estimated from Doppler shifts.

Astronomers often use *Pogson quantified magnitudes*, a logarithmic system to describe the incident flux in a certain frequency bandpass. There are many different variants, but the one that will appear most to physicists is the "absolute" system with

$$m_{AB} \equiv -2.5 \log f_\nu - 48.60 \quad . \tag{1.23}$$

The zero points of the various magnitude systems are arbitrary [2], but they all try, to some extent to preserve Ptolemy's scheme of 2000 years ago. The surface brightness μ (in dimensionless units of magnitude per square arcsecond) is then

$$\mu \equiv -2.5 \log I_\nu - 48.60 - 2.5 \log (206265)^2 \quad . \tag{1.24}$$

Observers will immediately recognize the last quantity in parentheses as the number of arcseconds in a radian (and the number of astronomical units in a parsec). As a matter of convention, the visible and near infrared portions of the electromagnetic spectrum are split up into a number spectral bands that have their origin in recipes for the glass that was used (by Harold Johnson) to construct filters. In order of increasing wavelength they are called U, B, V, R, I, J, H, and K where the first five letters stand for "ultraviolet", "blue", "visual", "red", and "infrared". When giving the luminosity or brightness of a source, one typically specifies a wavelength. For example, the background magnitude of the night sky in the B filter is roughly $\mu_B \sim 22.05$ $B_{mag}/arcsec^2$, corresponding to a luminosity surface density of $100 L_\odot/pc^2$.

Empirically, it is found that many galaxies obey simple "laws" that give the surface brightness as a function of radius from the galactic center. The best-known of these empirical relations are the *Freeman disk* and the *deVaucouleurs spheroid*:

$$I(r) = \begin{cases} I_0 \exp\left(\frac{r}{r_s}\right) & : \text{Freeman disk} \\ I_0 \exp\left[-7.67 \left(\frac{r}{r_e}\right)^{1/4}\right] & : \text{deVaucouleurs spheroid.} \end{cases} \tag{1.25}$$

For disks, the central intensity is such that at $r = 0$, the surface brightness is $\mu_0 \sim 21.5$ $B_{mag}/arcsec^2$. For spheroids, the effective radius, r_e, is defined so that half the light lies interior to it. At $r = r_e$ the surface brightness is typically $\mu_e \sim 22.5$ $B_{mag}/arcsec^2$. These are both very close to the surface brightness of the background night sky, which is made up primarily of zodiacal light and terrestrial airglow. This makes it relatively difficult to detect low surface brightness (LSB) galaxies, and those given to fretting (or quibbling) might worry that we were missing a large fraction of the universe

---

[2] This particular definition gives a magnitude for the star Vega of $m_{AB} \simeq 0$ at $5556 Å$

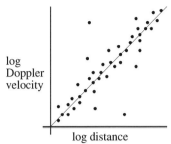

Figure 1.12: A schematic plot of redshift versus distance for a sample of galaxies. The slope is unity, implying a linear relation. Notice that a few points lie far from the mean. Real data is never pretty.

in failing to find such galaxies. Recent work by Blanton et al. using the Sloan Digital Sky Survey indicate that at most $\sim 5\%$ of the total luminosity density of the Universe might be lurking in galaxies too faint to detect against the surface brightness of the night sky.

In addition to these scaling laws that describe the surface brightness of a single galaxy, there are also a number of relationships that describe how galaxies of different luminosities are distributed throughout the Universe. Unfortunately, intrinsic luminosities cannot be determined without knowing the distance to a galaxy. Distances are difficult to measure. This would be a serious problem were it not for the Hubble relation, which tells us (to first order) that distance is proportional to Doppler shift, $D = v/H_0$ (see Figure 1.12). The Hubble constant, H_0, was poorly known until recently.

In the absence of a definitive measure of H_0, it was absorbed into many results, giving an overall map of the universe with a sliding scale. An example is *Schechter's Law*, an empirical expression for the "luminosity function" [# galaxies/unit volume with luminosity between L and $L + dL$],

$$\phi(L)dL = \phi^\star \left(\frac{L}{L^\star}\right)^\alpha \exp\left(-\frac{L}{L^\star}\right) d(L/L^\star). \tag{1.26}$$

Schechter's law works best for galaxies in the range $0.003 < L/L^\star < 3$ where $L^\star (\simeq 10^{10} L_\odot h^{-2})$ is determined from a fit giving $m_B^\star = -19.75 + 5\log h$, $\phi^\star = 2.0 \times 10^{-2} h^3 \text{Mpc}^{-3}$, and $\alpha = -1.09$. In each of these parameters, h is the dimensionless expansion factor of the Universe as a fraction of the canonical value for the Hubble constant[3]:

$$h \equiv \frac{H_0}{100 \text{ km/s/Mpc}} = \frac{v/D}{100 \text{ km/s/Mpc}}. \tag{1.27}$$

The luminosity function $\phi(L)$ has a form often seen in physical processes, a power law dependence with an exponential cutoff, shown schematically in Figure 1.13. Although $\phi(L)$ diverges for faint galaxies ($L \to 0$), the integrated luminosity is finite and actually analytic:

$$\int \phi(L) L dL = \Gamma(\alpha + 2)\phi^\star L^\star = (1 + \alpha)! \phi^\star L^\star. \tag{1.28}$$

Given the difficulty of measuring absolute distances and magnitudes of galaxies, it is useful to establish that one or another observable property of galaxies is correlated (the more strongly the better) with the intrinsic luminosity of a galaxy. The *Tully-Fisher Law* for spiral galaxies is one such

---

[3] The best measurements today give $h \simeq 0.7$, or equivalently, $H_0 \simeq (14 \text{Gyr})^{-1}$.

1.4. POTENTIAL THEORY

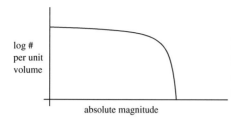

Figure 1.13: A schematic representation of the luminosity function for galaxies, cutting off sharply at the bright end and diverging slowly at the faint end.

empirical relation, between the circular velocity v_c and the total luminosity, L, of a galaxy, with

$$v_c \simeq 250 \text{ km/s} \left(\frac{L}{L^\star}\right)^{1/4} \qquad (1.29)$$

and the *Faber-Jackson Law* gives a similar relationship for the line-of-sight velocity dispersion σ_{los} for elliptical galaxie, :

$$\sigma_{los} \sim 220 \text{ km/s} \left(\frac{L}{L^\star}\right)^{1/4} . \qquad (1.30)$$

The *fundamental plane* relation for ellipticals is an extension of the Faber-Jackson relation to include a dependence upon surface brightness. It is usually expressed in terms of deVaucouleurs' effective radius r_e rather than luminosity,

$$r_e \propto \sigma_e^{1.49\pm0.05} I_e^{-0.75\pm0.01} . \qquad (1.31)$$

Here effective radius, r_e, depends upon distance, but the velocity dispersion and intensity do not.

Another two techinques for estimating galaxy distances involve the luminosity functions for globular clusters and planetary nebulae within galaxies. Yet another technique (the surface brightness fluctuation method) involves measuring the graininess of the observed surface brightnesses of elliptical galaxies due to the finite numbers of stars in each square arcsecond. The best technique requires patience and luck – observing a supernova within a galaxy.

1.4 Potential theory

The central result of Newton's gravitational theory is the *inverse square law* for the force between two point masses m_1 and m_2 separated by $\vec{r} = \vec{x}_1 - \vec{x}_2$:

$$\vec{F}(\vec{r}) = -\frac{Gm_1m_2}{r^2}\hat{r}. \tag{1.32}$$

Since the inverse square law can be added linearly for multiple point masses, we can write the force per unit mass caused by a continuum mass density distribution $\rho(\vec{x})$ as

$$\vec{F}(\vec{x}) = G \int \frac{\vec{x}' - \vec{x}}{|\vec{x}' - \vec{x}|^3} \rho(\vec{x}') d^3x', \tag{1.33}$$

giving the corresponding gravitational potential

$$\Phi(\vec{x}) = -G \int \frac{\rho(\vec{x}') d^3x'}{|\vec{x}' - \vec{x}|}, \tag{1.34}$$

(Continued on next page.)

from the force-potential relation for conservative forces, $\vec{F} = -\vec{\nabla}\Phi$. Integrating the Poisson equation (which can be derived by taking the divergence of the force equation) and applying the divergence theorem, we can derive Gauss' Theorem for localized mass distributions:

$$\int \vec{\nabla}\Phi \cdot d\vec{S} = 4\pi G \int \rho(\vec{x}) d^3 \vec{x} = 4\pi G M_{\text{enclosed}}. \tag{1.35}$$

The potential energy of a system (as distinct from the gravitational potential) is defined as the work done against gravitational forces in the assembly of a mass distribution. Calculation of the total work proceeds by examining the change in the gravitational potential $\delta\Phi(\vec{x})$, due to a change in the density $\delta\rho(\vec{x})$. After integration by parts and invocation of the divergence theorem, one finds

$$W = -\frac{1}{8\pi G}\int |\nabla\Phi(\vec{x})|^2 d^3\vec{x} = \frac{1}{2}\int \rho(\vec{x})\Phi(\vec{x})d^3\vec{x}, \tag{1.36}$$

where the minus sign indicates that negative work was done in assembling the mass distribution.

While these results have the advantage of being completely general, in practice the integrals can be solved analytically only for a few special cases. One of the most important simplifications result from spherical symmetry and the first two of *Newton's Theorems*:

a) A body inside a uniform density spherical shell experiences no net gravitational force from that shell.

b) The gravitational force on a body outside such a shell is the same as it would be if the shell's mass were concentrated at its center.

Both theorems can be proved through direct integration of equation (1.33) or by geometric arguments[4]. So for spherical symmetry, the potential can be divided into two pieces, that caused by the interior mass and that caused by the mass exterior to each radius r:

$$\Phi(r) = -4\pi G \left[\underbrace{\frac{1}{r}\int_0^r \rho(r')r'^2 dr'}_{\text{interior}} + \underbrace{\int_r^\infty \rho(r')r' dr'}_{\text{exterior}} \right]. \tag{1.37}$$

As we saw above in section (1.2), one result of spherical symmetry is the relation between the circular orbital velocity v_c and the potential $\Phi(r)$

$$v_c^2(r) = r\frac{d\Phi}{dr}. \tag{1.38}$$

The escape speed is defined as the velocity for which a particle has positive total energy (kinetic plus negative potential) and is thus not gravitationally bound to the system. For spherical symmetry, the escape speed v_e is a function only of radius

$$v_e^2(r) = -2\Phi(r), \tag{1.39}$$

---

[4]For proofs of these theorems as well as the *morthirabshearem*, see Binney and Tremaine.

where the convention adopted is that $\Phi(\infty) \to 0$. The Virial Theorem [see below, sec. (1.6)] gives another useful result relating the kinetic energy to the work done by gravity, averaged over an entire orbit:

$$\langle v^2 \rangle_{\text{orbit}} = \langle \vec{r} \cdot \vec{\nabla} \Phi \rangle_{\text{orbit}}. \tag{1.40}$$

Note that all three of the above equations have a "typical" or "characteristic" velocity squared on the left hand side and the gravitational potential on the right. In two of them the potential is differentiated with respect to length and then multiplied by a length that brings us back to the units of potential – velocity squared.

1.5 Density-potential pairs & orbits

Recall that the first step in constructing a galaxy is to pick a gravitational potential. Since differentiation is generally much more manageable than integration, it should be no surprise that we rely on Poisson's Equation to get the density distribution from the potential ($\Phi \to \rho$ via differentiation rather than $\rho \to \Phi$ via integration). We shall loosely speak of a density-potential pair being a "model" for a galaxy, but such models are incomplete. As we saw in Figure 1.3 there are many ways to populate a given density with different orbits. A complete description requires not just a physical density $\rho(\vec{x})$ but a phase space density $f(\vec{x}, \vec{p})$ (# stars/unit volume/unit momentum).

Different density-potential pairs are good for different purposes. Some have the virtue of simplicity or some special symmetry. Others permit one to write the phase space density in closed form, often a function of only a limited number of variables: $f = f(E)$, or $f = f(E, L^2)$ or $f = f(E, L_z)$, where E, L and L_z are respectively the total energy of the orbit, its angular momentum and its angular momentum around the z-axis. Yet others have the property that orbits are self-similar, meaning that an orbit at any radius can be scaled by an arbitrary factor producing another legitimate orbit. This is especially useful when generating orbit libraries.

Before proceeding we will want to tighten up our language a bit. We shall use *ellipsoid* to mean something whose equi-something contours are elliptical in cross-section but which generically have three unequal axes. A *spheroid* has also elliptical cross-sections but has two equal axes. *Oblate spheroids* have their long axes equal while *prolate spheroids* have their short axes equal.

We present the following list of density-potential pairs so that our reader is not taken by surprise when he encounters one in a talk or in the literature. He might even ask "Please remind me what special property of the something-something potential makes it appropriate for the problem at hand" and then say "ah yes" upon hearing the answer.

1.5.1 spherical potentials:

a) **point mass:** The Kepler potential with

$$\Phi(r) = -\frac{GM}{r}. \tag{1.4}$$

The energy of an object with mass μ is a function only of the elliptical orbit's semi-major axis

a:
$$E(a) = -\frac{GM\mu}{2a}. \tag{1.42}$$

b) **homogeneous sphere:** A constant density sphere of mass $M = \frac{4}{3}\pi\rho R^3$ gives orbits within the sphere with constant periods

$$T = \sqrt{\frac{3\pi}{G\rho}}. \tag{1.43}$$

Formulæ like this are common throoughout astrophysics. The typical dynamical time scale $T_{\rm dyn}$ must vary as $(G\rho)^{-1/2}$. The potential outside a homogeneous sphere is the same as that of the point mass $\Phi(r) = -GM/r$ while for $r < R$,

$$\Phi(r) = -2\pi G\rho\left(R^2 - \frac{r^2}{3}\right). \tag{1.44}$$

c) **isochrone:** This potential has the interesting property that the orbital period is a function only of the energy of the orbit. Recall that the Kepler potential has a similar degeneracy – the energy is a function only of the semi-major axis.

$$\Phi(r) = -\frac{GM}{b + \sqrt{b^2 + r^2}}, \tag{1.45}$$

giving a central density of

$$\rho_o = \frac{3M}{16\pi b^3} \tag{1.46}$$

d) **modified Hubble:** The modified Hubble law starts with a surface brightness profile that at one time was thought to be a close fit to those observed in for elliptical galaxies,

$$I(r) = \frac{2j_0 a}{1 + (R/a)^2}, \tag{1.47}$$

where j_0 is a central surface brightness and a is a scale length. It is a variant of a similar one called the *Hubble-Reynolds Law*. To first order it matches the profile for a non-singular isothermal sphere. The scale length plays the role of the core radius of the non-singular isothermal sphere. The associated analytic potential is somewhat cumbersome, but it is still of some use (as we shall see below) for clusters of galaxies.

e) **power-law density:** These models have simple power-law expressions for the density profile with

$$\rho(r) = \rho_o\left(\frac{r_o}{r}\right)^\alpha. \tag{1.48}$$

A special case of the power-law density potential is the *singular isothermal sphere*. It has a constant velocity dispersion and a density profile $\rho \propto r^{-2}$. This gives our old friend, the logarithmic potential described in Sec. (1.2). It has a flat rotation curve and infinite mass and radius. The logarithmic potentials produce self-similar orbits. There are therefore theoretical as well as practical reasons for choosing them.

1.5. DENSITY-POTENTIAL PAIRS & ORBITS

f) **Dehnen (with special cases Jaffe and Hernquist):** These have "cuspy" density profiles fo small r and a power law of the form $\rho \propto r^{-4}$ as $r \to \infty$. The phase-space distribution functio can be written in closed form as a function of energy and angular momentum, $f(E, L^2)$.

g) **King models:** Starting with the isothermal sphere, King introduced an energy cutoff to th distribution function that limits the size of the system. The associated *truncation radius* r_T often suggestively called a tidal radius.

h) **Navarro-Frenk-White:** This is an empirical fit to the observed density profiles in gravita tional N-body experiments,
$$\rho(r) = \frac{4\rho_s}{(r/r_s)(1 + r/r_s)^2} \quad , \tag{1.49}$$
where r_s is a scale length. It is a member of the larger family
$$\rho(r) = \frac{2^{3-\gamma} \rho_s}{(r/r_s)^\gamma (1 + r/r_s)^{(3-\gamma)}} \quad , \tag{1.50}$$
of which the $\gamma = 1.5$ case is called the *Moore profile*. Astronomy is not immune to fashion an these days the NFW and Moore profiles are quite fashionable, both as models for galaxy halo and for clusters of galaxies.

1.5.2 axisymmetric potentials

a) **Plummer-Kuzmin:** also known as "Toomre's model 1." In cylindrical coordinates $(R, z, \phi$ it has
$$\Phi(R, z) = \frac{-GM}{\sqrt{R^2 + (a + |z|)^2}} \quad , \tag{1.51}$$
giving a infinitesimally thin sheet of mass density at $z = 0$ with mass surface density
$$\Sigma(R) = \frac{aM}{2\pi (R^2 + a^2)^{3/2}} \quad . \tag{1.52}$$

b) **Miyamoto:** A combination of the Plummer-Kuzmin model and the spherical Plummer mode that is often used for globular clusters,
$$\Phi(R, z) = \frac{-GM}{\sqrt{R^2 + (a + \sqrt{b^2 + z^2})^2}} \quad . \tag{1.53}$$

c) **"cored" logarithmic:** A modified logarithmic potential that avoids a singularity at $R =$ and scales the z axis by the factor q so as to give a spheroidal potential, with $q < 1$ in th oblate case and $q > 1$ in the prolate case:
$$\Phi(R, z) = \frac{1}{2} v_o^2 \ln \left(R_c^2 + R^2 + \frac{z^2}{q^2} \right). \tag{1.54}$$

In the limit of vanishing core radius R_c, the equipotentials all have the same spheroidal shap and the equidensity contours likewise all have the same shape (though not spheroidal). Th orbits are therefore self-similar.

d) **Mestel disk:** As in Sec. (1.2), we write this in terms of spherical coordinates,

$$\Phi(r,\theta) = v_c^2 \left[\ln \frac{r}{R_c} + \ln \frac{1+|\cos\theta|}{2} \right]. \qquad (1.55)$$

e) **self-similar logarithmic:** As its name implies, the both the equipotential surfaces and the equidensity contours all have the same shape. It is separable in r and θ, with

$$\Phi(r,\theta) = v_o^2 \ln[(r \cdot Q(\theta)] \quad \text{and} \qquad (1.56)$$

$$\rho(r,\theta) = \rho_o \left(\frac{r_o}{r}\right)^2 S(\theta), \qquad (1.57)$$

where $Q(\theta)$ and $S(\theta)$ are arbitrary functions for the potential and density (but of course related to each other via Poisson's equation). The spheroidal logarithmic potentials describe above are special cases of this.

f) **homoeoids:** A homoeoid is a uniform density shell of finite thickness whose inner and outer surfaces are similar spheroids. A thin homoeoid is the limiting case where the inner and outer shells approach each other. Thin homoeoids have the remarkable property their external equipotentials are spheroids that are confocal with the thin homoeoid, while their internal potentials are constant. This can be used to generate the potentials of arbitrary spheroidal mass distributions. Many of the results for spheroidal systems are most simply expressed in *spheroidal coordinates*. If one draws nested confocal spheroids, there is a corresponding family of confocal hyperboloids of revolution. One coordinate is constant on the ellipses (playing the role of the radius in spherical coordinates) and the other is constant on the hyperboloids of revolution, playing the role of the polar angle.

1.5.3 non-axisymmetric potentials and the multipole expansion

The logarithmic potential and homoeoid above are straightforwardly generalized to non-axisymmetric cases. In the logarithmic case of the potential, the density and the functions Q and S are then functions of r, θ and ϕ. The homoeoids are by extension ellipsoidal rather than spheroidal.

For more general three-dimensional systems, last resort is to expand the potential in spherical harmonics:

$$\Phi(r,\theta,\phi) = -4\pi G \sum \frac{Y_{lm}(\theta,\phi)}{2l+1} \left[\frac{1}{r^{l+1}} \int_0^r \rho_{lm}(a) a^{l+2} da + r^l \int_r^\infty \frac{\rho_{lm}(a)}{a^{l-1}} da \right], \qquad (1.58)$$

where Y_{lm} are the spherical Legendre functions and

$$\rho_{lm}(a) = \int Y_{lm}^*(\theta,\phi) \rho(a,\theta,\phi) d\Omega. \qquad (1.59)$$

The multipole expansion method is frequently used in N-body simulations because of the computationally efficient methods for solving the Laplacian with spherical harmonics.

1.6 Orbits in spherical potentials

Our method for building galaxy models (often called "Schwarzschild's method," after Martin Schwarzson of the general relativist Karl) requires that we construct an orbit library in which we store the "footprints" or probability densities of a representative set of orbits. These are then superposed to construct the luminosity distribution. Since orbits play such a major role, we discuss them in some detail

We begin by calculating orbits for the simplest systems mentioned above, those with spherically symmetric potentials $\Phi = \Phi(r)$. It is easy to get lost in the details of the calculation, so we start with an outline of the procedure we will follow:

a) Discover that the vector angular momentum is conserved and reduce the dimensionality of the problem to motion in a plane.

b) Split the vector equation of motion into two scalar equations of motion.

c) Change the independent variable from time to angle.

d) Change the dependent variable from radius to inverse radius.

e) Multiply by the derivative of inverse radius with respect to angle and integrate, giving a differential equation for the square of the derivative of inverse radius with respect to angle. The constant of integration is proportional to the energy per unit mass divided by the angular momentum squared.

f) Solve the differential equation for inverse radius as a function of angle.

The experience of seeing this for the first time is not unlike that of an undergraduate being led from one side of the campus to the other via basements and tunnels.

We begin with the central force equation (where we adopt the convention of expressing force $\vec{F}$, energy E, and angular momentum $\vec{L}$ per unit mass)

$$\frac{d^2}{dt^2}\vec{r} = -\vec{\nabla}\Phi(r) = \vec{F} = F(r)\hat{e}_r. \tag{1.60}$$

Combined with the angular momentum definition

$$\vec{L} \equiv \vec{r} \times \vec{v}, \tag{1.61}$$

gives the conservation of angular momentum in central-force systems:

$$\frac{d}{dt}\vec{L} = \frac{d}{dt}\left(\vec{r} \times \frac{d\vec{r}}{dt}\right) = \frac{d\vec{r}}{dt} \times \frac{d\vec{r}}{dt} + \vec{r} \times \frac{d^2\vec{r}}{dt^2} = 0, \tag{1.62}$$

so we immediately see that the orbit will be constrained to a plane perpendicular to $\vec{L}$. Within this plane we adopt coordinates (r, ψ) with

$$L = r^2\frac{d\psi}{dt}. \tag{1.63}$$

The equations of motion in this new coordinate system are

$$\vec{r} = r\hat{e}_r$$
$$\frac{d}{dt}\vec{r} = \dot{r}\hat{e}_r + r\dot{\psi}\hat{e}_\psi$$
$$\frac{d^2}{dt^2}\vec{r} = (\ddot{r} - r\dot{\psi}^2)\hat{e}_r + (2\dot{r}\dot{\psi} + r\ddot{\psi})\hat{e}_\psi, \qquad (1.64)$$

where we have used the relations

$$\frac{d}{dt}\hat{e}_r = \dot{\psi}\hat{e}_\psi \text{ and } \frac{d}{dt}\hat{e}_\psi = -\dot{\psi}\hat{e}_r. \qquad (1.65)$$

Separating the centeral force equation (1.60) into two scalar equations of motion gives

$$\ddot{r} - r\dot{\psi}^2 = F(r) \qquad (1.66)$$

and

$$2\dot{r}\dot{\psi} + r\ddot{\psi} = 0. \qquad (1.67)$$

We change the both the dependent and independent variables in the radial equation, to ψ and $u \equiv 1/r$ respectively. We then have

$$\frac{d}{dt} = \frac{L}{r^2}\frac{d}{d\psi} \qquad (1.68)$$

from equation (1.63) and

$$\frac{1}{r^2}\frac{d}{d\psi}r = -\frac{d}{d\psi}u. \qquad (1.69)$$

The radial equation (1.66) becomes

$$\frac{d^2u}{d\psi^2} + u = -\frac{F(u)}{L^2 u^2}. \qquad (1.70)$$

We assume that $F(r)$ can be derived from a potential $F(r) = -\nabla\Phi$. We then use the common trick of turning a force equation into an energy equation by multiplying by a first derivative of position (in our case $\frac{du}{d\psi}$) and integrating with respect to the timelike variable ψ giving

$$\left(\frac{du}{d\psi}\right)^2 + u^2 + \frac{2\Phi}{L^2} = \frac{2E}{L^2}, \qquad (1.71)$$

where E is the conserved energy per unit mass, expressed in our original coordinates as

$$E = \frac{1}{2}\left(\frac{dr}{dt}\right)^2 + \frac{1}{2}r^2\left(\frac{d\psi}{dt}\right)^2 + \Phi(r). \qquad (1.72)$$

We can solve for the turning points in the orbit (where $\dot{r} = 0$) by setting $E = \Phi_{\text{eff}}$, where $\Phi_{\text{eff}}(r)$ is the effective potential

$$\Phi_{\text{eff}} = \frac{1}{2}\frac{L^2}{r^2} + \Phi(r). \qquad (1.73)$$

To procede further we must choose a potential. There are a wide variety of potentials used in the study of galaxies. Most of these require numerical integration of orbits. Here we consider some special cases which can be solved analytically and which serve to make interesting points about orbits.

1.6. ORBITS IN SPHERICAL POTENTIALS

a) **Harmonic Potential**: This is just like a spherically symmetric harmonic oscillator, giving closed elliptical orbits centered on the origin,

$$\Phi(r) = \frac{1}{2}\Omega^2 r^2. \tag{1.74}$$

b) **Kepler Potential**: This is the gravitational potential for a single point mass,

$$\Phi(r) = -\frac{GM}{r} \quad . \tag{1.75}$$

Taking the derivative of this potential with respect to r and substituting into our equation for the second derivative of inverse radius with respect to angle, equation (1.70) gives

$$\frac{d^2 u}{d\psi^2} + u = \frac{GM}{L^2} \quad . \tag{1.76}$$

which is just the equation for a simple harmonic in u with

$$u = C\cos(\psi - \psi_0) + \frac{GM}{L^2} \quad . \tag{1.77}$$

Rearranging terms we have

$$r = \frac{a(1-e^2)}{1 + e\cos(\psi - \psi_0)} \quad , \tag{1.78}$$

where

$$a \equiv \frac{L^2}{GM(1-e^2)} \quad , \tag{1.79}$$

and e is the eccentricity. Note that there are three constants in the above equations, a, e and ψ_0. But there are two more hidden constants, the angles giving the orientation of the orbital plane. Thus the orbit traces a one dimensional locus in six dimensional phase space. We shall loosely call such constants which reduce the dimensionality of the allowed phase space *integrals of the motion*.

c) **Post-Newtonian Kepler Potential**: To lowest order the effects of general relativity can be approximated by adding a "post-Newtonian" correction to the Newtonian potential,

$$\Phi(r) = -\frac{GM}{r}\left(1 + \frac{k}{r}\right) \quad , \tag{1.80}$$

where $k = 2GM/c^2$ has units of length. Substituting this into our general equation for the spherically symmetric case (1.71) we find

$$\frac{d^2 u}{d\psi^2} + u\left(1 - \frac{2GMk}{L^2}\right) = \frac{GM}{L^2} \quad . \tag{1.81}$$

This *again* gives simple harmonic motion in the radial direction, but the radial frequency is no longer unity. The radial and azimuthal oscillations now have different periods, and the orbits are no longer closed ellipses. The periastron precesses. One of the early successes of the theory of general relativity was its successful explanation of the precession of the perihelion of Mercury's orbit. For larger k the orbit looks like the "rosette" shown in Figure 1.4.

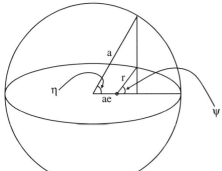

Figure 1.14: An elliptical orbit with semi-major axis a and eccentricity e. The eccentric anomaly η is defined by circumscribing a circle.

d) **Witta-Paczyński Potential:** This has the functional form

$$\Phi(r) = -\frac{GM}{r - \frac{2GM}{c^2 r}}, \tag{1.82}$$

which reduces to the post-Newtonian approximation in the limit of large r. It also has the property of having an innermost stable circular orbit (ISCO) that coincides with the innermost stable orbit of a Schwarzschild black hole. It allows approximate Newtonian calculations of the dynamics of accretion disks without the apparatus of general relativity.

1.6.1 digression on the Kepler problem

The solution provided to the Kepler problem above is mathematically correct and practically useless. Observations give position or velocity as a function of time, not angle. Unfortunately there is no closed form expression for the position as a function of time. But position and time can be given parametrically in terms of an angle η called the *eccentric anomaly*. The angle ψ is called the *true anomaly*. Both angles are shown in Figure 1.14. Thus parameterized we have

$$r = a(1 - e\cos\eta) \quad \text{and} \tag{1.83}$$

$$t = \frac{P}{2\pi}(\eta - e\sin\eta), \tag{1.84}$$

where P is the orbital period given by Kepler's third law

$$P = 2\pi\sqrt{\frac{a^3}{GM}}. \tag{1.85}$$

Once can solve for the true anomaly in term of the eccentric anomaly, and *vice versa* using

$$\tan\frac{\psi}{2} = \left(\frac{1+e}{1-e}\right)^{1/2}\tan\frac{\eta}{2}. \tag{1.86}$$

We will see this parametric solution to Kepler's potential again when studying the gravitational expansion and collapse of the Universe.

1.7. ORBITS IN AXISYMMETRIC POTENTIALS

1.6.2 digression on the virial theorem

So as not to interrupt our narrative thread we postponed a proof of the virial theorem, equation (1.40)

$$\langle v^2 \rangle_{\text{orbit}} = \langle \vec{r} \cdot \vec{\nabla}\Phi \rangle_{\text{orbit}} \quad .$$

Having now given further attention to orbits we proceed with our proof. Taking the dot product equation (1.60) with $\vec{r}$ and averaging over time gives

$$\left\langle \vec{r} \cdot \frac{d^2\vec{r}}{dt^2} \right\rangle = \lim_{T \to \infty} \frac{1}{T} \int_0^T \vec{r} \cdot \frac{d^2\vec{r}}{dt^2} dt = \lim_{T \to \infty} \frac{1}{T} \left[\vec{r} \cdot \frac{d\vec{r}}{dt} \Big|_0^T - \int_0^T \left(\frac{d\vec{r}}{dt}\right)^2 dt \right]. \tag{1.8}$$

For bounded orbits, the long-time average of $\vec{r} \cdot \frac{d\vec{r}}{dt}$ is zero, so we are left with

$$-\left\langle \vec{r} \cdot \frac{d^2\vec{r}}{dt^2} \right\rangle_{\text{time}} = \left\langle \left(\dot{\vec{r}}\right)^2 \right\rangle_{\text{time}} = \left\langle \vec{r} \cdot \vec{\nabla}\Phi \right\rangle_{\text{time}} = \left\langle v_c^2 \right\rangle_{\text{time}} \tag{1.8}$$

where the last equality holds only for cylindrically symmetric potentials. These results can be very helpful in calculating distributions of orbits in phase space, but one must be careful not to abuse the Virial Theorem. One may, for example, replace the *time* average in equation (1.88) with a *volume* average, but only if all stars and orbits are inclueded. The result does not apply, in particular, if the average is taken over only a finite region of position space.

1.7 Orbits in axisymmetric potentials

We have seen above that empirically, most galaxies are *not* spherically symmetric, but rather have flattened disks or ellipsoidal shapes. For many of these more general potentials, azimuthal symmetry and also up-down symmetry ($z \to -z$) are still fine approximations. Thus we can write the potential as $\Phi(R, \phi, z) = \Phi(R, |z|)$. From the formula for the gradient in cylindrical coordinates, we get the three pieces of the vector equation $\ddot{\vec{r}} = -\vec{\nabla}\Phi$,

$$\begin{aligned} \hat{\mathbf{e}}_R &: \ddot{R} - R\dot{\phi}^2 = -\frac{\partial \Phi}{\partial R} \\ \hat{\mathbf{e}}_\phi &: \frac{d}{dt}(R^2 \dot{\phi}) = 0 \\ \hat{\mathbf{e}}_z &: \ddot{z} = -\frac{\partial \Phi}{\partial z}. \end{aligned} \quad (1.89)$$

Just as in the spherically symmetric case, we can define an effective potential Φ_{eff} for a given angular momentum about the z-axis, $L_z = R\dot{\phi}^2$,

$$\Phi_{\text{eff}} = \Phi + \frac{L_z^2}{2R^2} \quad . \quad (1.90)$$

The equation of motion in the $\hat{\mathbf{e}}_R$ direction is

$$\ddot{R} = -\frac{\partial \Phi_{\text{eff}}}{\partial R} \quad , \quad (1.91)$$

(Continued on next page.)

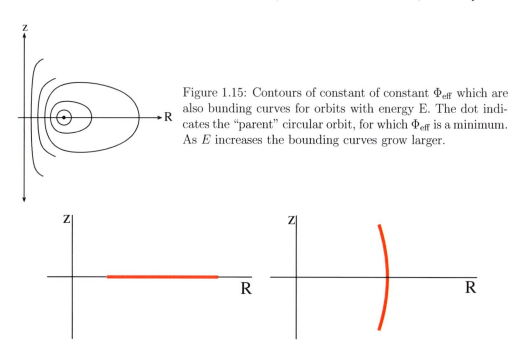

Figure 1.15: Contours of constant of constant Φ_{eff} which are also bunding curves for orbits with energy E. The dot indicates the "parent" circular orbit, for which Φ_{eff} is a minimum. As E increases the bounding curves grow larger.

Figure 1.16: Left: A circular orbit has been perturbed in the R direction. Right: A circular orbit has been perturbed in the z direction.

the equation of motion in the z direction is

$$\ddot{z} = -\frac{\partial \Phi_{\text{eff}}}{\partial z}, \tag{1.92}$$

and the energy equation is

$$E = \frac{1}{2}(\dot{R}^2 + \dot{z}^2) + \Phi_{\text{eff}}. \tag{1.93}$$

If we draw curves of constant effective potential (Fig. 1.15), we see that for a given L_z the minimum effective energy is at a point R_g on the R axis. A particle with exactly this minimum energy will be in circular orbit, with orbital frequency Ω such that

$$\Omega^2 = \frac{1}{R}\frac{\partial \Phi}{\partial R}. \tag{1.94}$$

We can expand the effective potential in the vicinity of this point,

$$\Phi_{\text{eff}} \approx \Phi(R_g, 0) + \frac{1}{2}\left(\frac{\partial^2 \Phi_{\text{eff}}}{\partial R^2}\right)_{(R_g,0)} x^2 + \frac{1}{2}\left(\frac{\partial^2 \Phi_{\text{eff}}}{\partial z^2}\right)_{(R_g,0)} z^2, \tag{1.95}$$

where we have defined a radial displacement, $x \equiv R - R_g \ll R_g$.

If a test particle in circular orbit is given a small perturbation in the radial direction, (Fig. 1.16) it obeys the equation of motion

$$\ddot{x} + \kappa^2 x = 0 \quad \text{where} \tag{1.96}$$

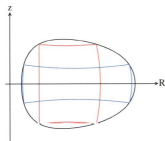

Figure 1.17: Bounding curves for two orbits with the same values of E and L_z. Evidently there is some third conserved quantity that makes for distinct bounding curves.

$$\kappa^2 \equiv \left(\frac{\partial^2 \Phi_{\text{eff}}}{\partial R^2}\right)_{(R_g,0)} \tag{1.97}$$

defines the *epicyclic frequency*, κ.

Similarly giving a particle given a small perturbation perpendicular to the orbital plane (Fig. 1.16) will obey the equation of motion

$$\ddot{z} + \nu^2 z = 0 \quad \text{where,} \tag{1.98}$$

$$\nu^2 \equiv \left(\frac{\partial^2 \Phi_{\text{eff}}}{\partial z^2}\right)_{(R_g,0)} \tag{1.99}$$

defines the vertical frequency ν.

As the amplitudes of the the perturbations increase the expansion above is no longer strictly correct, but the character of the two orbits remains the same. In one case the particle oscillates within the orbital plane, while in the other the particle traces out an arc in the (R, z), *meridional plane*. Notice that the epicyclic, vertical and orbital frequencies need not be the same – in general they *are* different. Only for the Kepler potential and for the spherical harmonic potential are they identical.

More generally, a particle given an arbitrary perturbation must be contained within the contour $\Phi_{\text{eff}} = E$ in Figure 1.15. But when a circular orbit is given an arbitrary perturbation, the orbit does not fill the allowed region. Rather it produces a Lissajou figure contained within a box-shaped region. (Fig. 1.17). Notice that two different box-shaped regions can have the same bounding energy. This suggests that some additional conserved quantity, another integral of the motion (beyond energy and z-axis angular momentum), denies it access to the full allowed region.

To study the dimensionality of these orbits, one constructs a Poincare *surface of section* which reduces the six-dimensional phase space to something more manageable. The construction proceeds as follows:

a) From azimuthal symmetry, ignore ϕ and $\dot{\phi}$, giving 4-D trajectories in $(R, \dot{R}, z, \dot{z})$.

b) Use conservation of energy to give $\dot{z} = \dot{z}(R, \dot{R}, z, E)$

c) Take a cross-section of the orbits, plotting a point on the $(R, \dot{R})$ plane every time the orbit crosses the plane defined by $z = 0$ with $\dot{z} > 0$.

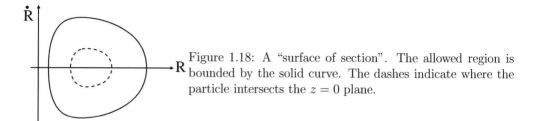

Figure 1.18: A "surface of section". The allowed region is bounded by the solid curve. The dashes indicate where the particle intersects the $z=0$ plane.

d) The confinement of trajectories to 1-D invariant curves in the $(R, \dot R)$ plane suggests they will occupy 2-D surfaces in $(R, \dot R, z, \dot z)$ space and 3-D regions within the full 6-D phase space and thus the existence of an additional integral of the motion.

An example of such a Poincare surface of section is shown in Figure 1.18.

For potentials that are nearly spherical, the third integral is something like, but not exactly the total angular momentum. For orbits that stay close to the equatorial plane, the third integral is something like, but not exactly, the energy in the z direction. More generally, while one may demonstrate the existence of a third integral using Poincare's method, it cannot be written in closed form[5].

1.7.1 nearly circular orbits: the epicyclic approximation

For many axisymmetric potentials, we are interested primarily in trajectories that are nearly circular with an average orbital frequency $\Omega(R)$ at each radius R. This circular orbit acts as a *guiding center* for the perturbed orbit which we want to study. Near this equilibrium orbit, we can expand the effective potential around the guiding radius R_g and the equatorial plane $z = 0$,

$$\Phi_{\text{eff}} \approx \Phi(R_g, 0) + \frac{1}{2}\left(\frac{\partial^2 \Phi_{\text{eff}}}{\partial R^2}\right)_{(R_g,0)} x^2 + \frac{1}{2}\left(\frac{\partial^2 \Phi_{\text{eff}}}{\partial z^2}\right)_{(R_g,0)} z^2 \quad . \tag{1.100}$$

We pay attention here only to motion in the plane, which decouples from the vertical motion, and henceforth take z to be identically zero. As before we take the x direction to be radial and the y direction to be positive in the direction of rotation. As noted in the previous section, small osciallations are governed by the harmonic equation

$$\ddot x + \kappa^2 x = 0 \tag{1.101}$$

and κ^2, defined by equation (1.97), is then

$$\kappa^2 = \left(\frac{\partial^2 \Phi}{\partial R^2}\right)_{R_g} + \frac{3L_z^2}{R_g^4} = \left(R\frac{d}{dR}\Omega^2 + 4\Omega^2\right)_{R_g} . \tag{1.102}$$

---

[5] An interesting counterexample is the tri-axial Stäckel potential where all three integrals of motion *can* be written in closed form.

1.7. ORBITS IN AXISYMMETRIC POTENTIALS

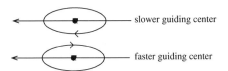

Figure 1.19: Stars executing epicycles around guiding centers or circular orbits. The inner orbit has a higher orbital frequency.

Recall from equation (1.15) the relation between circular velocity and potential gives the orbital frequency as

$$\Omega^2 = \frac{1}{R}\frac{\partial \Phi}{\partial R} = \frac{L_z^2}{R^4}. \tag{1.103}$$

The solutions to (1.101) are harmonic oscillations of the form

$$x(t) = X\cos(\kappa t + \psi), \tag{1.104}$$

where ψ is a phase determined by initial conditions.

But what is happening in the ϕ direction as the particle excutes radial oscillations? The angular position in the orbit can be solved for using conservation of angular momentum and the fact that that $x \ll R_g$,

$$\dot\phi = \frac{L_z}{R^2} = \frac{L_z}{R^2\left(1+\frac{x}{R_g}\right)^2} \approx \Omega_g\left(1 - \frac{2x}{R_g}\right). \tag{1.105}$$

From here we integrate to get

$$\phi(t) = \Omega_g t + \phi_o - \frac{2\Omega_g}{\kappa R_g}X\sin(\kappa t + \psi). \tag{1.106}$$

It is useful to define y as the distance away from the guiding circular orbit in the angular direction:

$$y \equiv R_g(\phi - \phi_g) = -\frac{2\Omega_g}{\kappa}X\sin(\kappa t + \psi) = Y\sin(\kappa t + \psi). \tag{1.107}$$

As both the x and y coordinates execute harmonic oscillations with the same frequency, κ, the trajectories trace out ellipses, called *epicycles* around the guiding center[6]. A few helpful relations can be derived without too much effort:

$$\frac{1}{2} < \frac{X}{Y} = \frac{\kappa}{2\Omega_g} < 1, \tag{1.108}$$

where the lower bound (1/2) corresponds to a Keplerian potential and the upper bound (1) corresponds to a harmonic potential. Averaging over an epicycle ($\kappa t = 0 \to 2\pi$), we find

$$\frac{\langle \dot y^2\rangle_{\text{orbit}}}{\langle \dot x^2\rangle_{\text{orbit}}} = \frac{4\Omega_g^2}{\kappa^2}. \tag{1.109}$$

---

[6]Ptolemy is unjustly ridiculed for having introduced epicycles to explain planetary motions 2000 years ago. Epicycles do indeed improve upon circular motion. What Ptolemy failed to realize was that for planetary motion these epicycles are ellipses rather than circles.

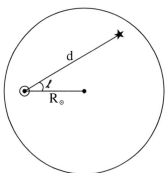

Figure 1.20: A star at distance d from the sun observed in the plane of the Milky Way at galactic longitude l. The sun is located a distance $R_\odot$ from the origin at the center of the galaxy.

However, if one takes the average over a volume of space encompassing many orbits instead of just averaging over a single orbit (something we postpone discussing until our treatment of Jeans equations), one finds the exact opposite:

$$\frac{\langle \dot{y}^2 \rangle_{\rm vol}}{\langle \dot{x}^2 \rangle_{\rm vol}} = \frac{\kappa^2}{4\Omega_g^2}. \tag{1.110}$$

A qualitative understanding for this can be seen in Figure 1.19, where the outer, slower guiding center overlaps with a faster inner guiding center but their epicyclic velocities are in opposite directions.

In the early part of this century, Jan Oort interpreted nearby stellar radial velocity and transverse proper motion measurements as the result of differential rotation, giving Ω as a function of R. From the point of view of an observer moving on a circular orbit at distance R from the center of the galaxy, a star at a distance d and galactic longitude l (see Figure 1.20) will have a line-of-sight radial velocity given by

$$v_{\rm los} = d(A \sin 2l) \tag{1.111}$$

and a transverse velocity of

$$v_T = d(A \cos 2l + B), \tag{1.112}$$

which is actually measured in terms of a proper motion (typically fractions of an arcsecond per yr)

$$\mu = v_T/d = A \cos 2l + B. \tag{1.113}$$

Here A and B are the *Oort constants* defined as

$$A \equiv -\frac{1}{2} R \frac{d\Omega}{dR} \tag{1.114}$$

and

$$B \equiv -(\frac{1}{2} R \frac{d\Omega}{dR} + \Omega). \tag{1.115}$$

From our definitions of the epicyclic frequency, equation (1.97) and Oort's constants we derive

$$\kappa^2 = -4B(A - B). \tag{1.116}$$

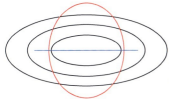

Figure 1.21: The black ellipses are equipotentials. The blue horizontal line is the radial orbit that is the parent of the box orbits. The red ellipse is the closed orbit that is the parent of the tube orbits.

1.8 2D non-axisymmetric potentials

For orbits in a non-axisymmetric potential, no component of the angular momentum is conserved. There are nonetheless two distinct families of orbits that contribute to the shape of the galaxy. The families are generated by stable, closed *parent orbits*.

Starting at any point on the long axis of the galaxy, one can launch particles perpendicular to that axis. In general these will generate rosettes that do not close upon themselves. But for exactly one value of velocity at every position on the long axis there is an orbit which returns to that point on the axis with the same velocity – it closes on itself, producing an orbit that is roughly elliptical (see Figure 1.21). Small perturbations to this orbit produce orbits that fill a narrow ribbon straddling this orbit. Such orbits are called *tube orbits*. As the pertubations increase, the width of the ribbon increases, but the sign of the angular momentum of the orbit does not change.

A second family has a parent that oscillates back and forth on the long axis. Small perturbations to this parent produces orbits that fill a bow or box shaped region along the long axis. Bigger perturbations produce wider bows and boxes. But the angular momentum is relatively small and changes sign frequently. These are called *box orbits*.

Box orbits (sometimes called *centrophilic*) are characterized by an angular momentum that changes sign, while tube orbits (*centrophobic*) always have L_z that does not change sign. Box orbits *must* be present for a disk to be non-axisymmetric, since the tube orbits are elongated counter to the elongation of the potential.

In addition to these major families, there are also minor families. The parent closed orbits for these families are characterized by the stars coming to a complete halt and reversing direction. The names for these families are particularly apt: banana orbits and fish orbits look like bananas and fishes. Small perturbations from the parents produce similarly shaped orbits, but larger perturbations produce boxes and tubes.

A further complication can be added if the non-axisymmetric potential is allowed to tumble. Barred galaxies are thought to be such tumbling non-axisymmetric shapes.

1.8.1 3D non-axisymmetric potentials

The most straightforward case of a 3D non-axisymmetric potentials would be one that was ellipsoidal with three unequal axes. Such potentials have three major families of orbits. There are *short-axis tubes* whose angular momentum is roughly aligned with the short axis, and *long-axis tubes* whose angular momentum is roughly aligned with the long axis. The projection of the angular momentum along these axes does not change sign. There are also box orbits, for which the projection of the angular momentum along each of the three axes does change sign. The tube orbits are generically torii. The box orbits are quite boxy.

Stäckel potentials are remarkable in being triaxial but admitting for closed form descriptions of its orbits in terms of confocal ellipsoidal coordinates. Staëkel potentials are useful for studying the qualitative nature of orbits but less useful for constructing realistic models.

There is overwhelming evidence for central mass concentrations at the centers of galaxies that are thought to be black holes. These produce cusps at centers of potentials that destroy box orbits. It may be that galaxies with large central black holes cannot be triaxial.

As with 2D non-axisymmetric potentials, 3D potentials may tumble, adding another degree of complication.

1.9 The collisionless Boltzmann equation and Jeans' equations

We have emphasized Schwarzschild's view of a galaxy as a superposition of orbits. He prescribes a scheme that is straightforward in principle but difficult in practice. An alternate view of galaxies is as a system of particles in six dimensional *phase space*. The galaxy is then instantaneously described as a distribution function $f(\vec{x}, \vec{v})$ over the phase space.

Since Schwarschild's approach assumes a time independent potential, it guarantees that a set of orbits that reproduces that potential will do so for all time. By contrast a phase space density that reproduces the density and potential of a galaxy at one instant will not in general reproduce itself at later (or earlier times). The time evolution of the phase space density is governed by a a 6-dimensional equation of continuity that is analogous to the familiar 3-dimensional equation of continuity of fluid mechanics. Each point in phase space is described by a 6-D vector $\vec{w} = (\vec{x}, \vec{v})$. The equation of continuity is then

$$\frac{\partial f}{\partial t} + \sum_{\alpha=1}^{6} \frac{\partial}{\partial w_\alpha}(f \dot{w}_\alpha) = 0. \tag{1.117}$$

We find that

$$\sum_{\alpha=1}^{6} \frac{\partial \dot{w}_\alpha}{\partial w_\alpha} = \sum_{i=1}^{3}\left(\frac{\partial v_i}{\partial x_i} + \frac{\partial \dot{v}_i}{\partial v_i}\right) = 0, \tag{1.118}$$

where the first part of the sum is zero because the velocities are necessarily not explicit functions of position (hence six dimensions in phase space and not fewer). The equation of motion tells us that $\dot{v}_i = -\frac{\partial}{\partial x_i}\Phi(\vec{x})$ and thus $\dot{v}_i$ is a function of position only. Substituting back into our 6-dimensional equation of continuity we get the collisionless Boltzmann equation (CBE):

$$\frac{\partial f}{\partial t} + \vec{v} \cdot \vec{\nabla} f - \vec{\nabla} \Phi \cdot \frac{\partial f}{\partial \vec{v}} = 0. \tag{1.119}$$

It collisionless in that particles do not make instantaneous jumps in $\vec{x}$ or $\vec{v}$, a consequence of a potential Φ that is smooth in space and time. The CBE can be used simultaneously for many different

1.9. THE COLLISIONLESS BOLTZMANN EQUATION AND JEANS' EQUATIONS

species in a galaxy, each with its own distribution function. This is important in implementing the fourth step of galaxy construction: calculating the distribution of light produced by all the different types of stars in the galaxy.

The phase space density $f(\vec{x}, \vec{v})$ is far too unwieldy to produce a useful description of a galaxy. But if the dependence of the phase space density upon velocity is relatively smooth and free of singularities, one can collapse the 6-dimensional phase space density into set of functions of 3-dimensional position by taking moments of the velocities. The zeroth moment is just the number denisity, $\nu(\vec{x})$, give by

$$\nu(\vec{x}) \equiv \int f(\vec{x}, \vec{v}) d^3v \quad . \tag{1.120}$$

For each of three velocity components the first moment gives a mean velocity,

$$\bar{v}_i(\vec{x}) \equiv \frac{1}{\nu(\vec{x})} \int v_i f(\vec{x}, \vec{v}) d^3v \quad . \tag{1.121}$$

One can likewise define higher order moments with combinations of powers of the three velocity components. The second moments give a quantity related to the velocity dispersion tensor, σ_{ij}^2,

$$\overline{v_i v_j}(\vec{x}) \equiv \frac{1}{\nu(\vec{x})} \int v_i v_j f(\vec{x}, \vec{v}) d^3v = \sigma_{ij}^2 + \bar{v}_i \bar{v}_j \quad . \tag{1.122}$$

Ideally the velocity distribution functions are not cuspy and so are reasonably well described by the low order moments. There is at least some observational support for this. A density and a set of low order moments may therefore give a reasonably complete description of a galaxy.

By multiplying the CBE by powers of the velocity components, and integrating over velocity space we obtain a series of differential equations for the various velocity moments. The zeroth moment of the CBE yields

$$\int \frac{\partial f}{\partial t} d^3v + \int v_i \frac{\partial f}{\partial x_i} d^3v - \frac{\partial \Phi}{\partial x_i} \int \frac{\partial f}{\partial v_i} d^3v \quad . \tag{1.123}$$

Note that we have dropped the summation signs and adopted the implicit summation convention, in our case over i. We can eliminate the last term term by application of the divergence theorem,

$$\int \vec{g} \cdot \vec{\nabla}_v f d^3v = \int f \vec{g} \cdot d\vec{S} - \int f \vec{\nabla}_v \cdot \vec{g} d^3v \quad . \tag{1.124}$$

In the present case $g = \vec{\nabla}\Phi(\vec{x})$ and is not a function of $\vec{v}$ and so may move freely inside and outside the integral. We have $\vec{\nabla}_v \cdot \vec{g} = 0$ and the surface integral goes to zero if the phase space density goes to zero at infinity. Incorporating our definitions of number density and mean velocity we have

$$\frac{\partial}{\partial t} \nu + \frac{\partial}{\partial x_i}(\nu \bar{v}_i) = 0, \tag{1.125}$$

which looks just like a standard 3-D continuity equation. This should come as no surprise since to derive it we just collapsed the 6-D continuity equation by integrating over all velocity.

The first moment of the CBE is found by multiplying by v_j and integrating over velocity. Again taking the spatial and temporal derivatives outside the velocity integrals, we get

$$\frac{\partial}{\partial t} \int v_j f d^3v + \frac{\partial}{\partial x_i} \int v_j v_i f d^3v - \frac{\partial \Phi}{\partial x_i} \int v_j \frac{\partial f}{\partial v_i} d^3v = 0 \quad . \tag{1.126}$$

Integrating by parts and expressing our results in terms of our average velocities, we have

$$\frac{\partial}{\partial t}(\nu \bar{v}_j) + \frac{\partial}{\partial x_i}(\nu \overline{v_i v_j}) + \frac{\partial \Phi}{\partial x_i} \int f \frac{\partial v_j}{\partial v_i} d^3\vec{v} = 0 \ . \tag{1.127}$$

In an orthogonal coordinate system, $\frac{\partial v_j}{\partial v_i} = \delta_{ij}$ so the last term on the left hand side becomes $\nu \delta_{ij}$. Applying the product rule and our continuity equation to the first term we get

$$\nu \frac{\partial \bar{v}_j}{\partial t} - \bar{v}_j \frac{\partial}{\partial x_i}(\nu \bar{v}_j) + \frac{\partial}{\partial x_i}[\nu(\sigma_{ij}^2 + \bar{v}_i \bar{v}_j)] = -\nu \bar{v}_i \frac{\partial \Phi}{\partial x_j} \ , \tag{1.128}$$

where we have made use of the relation between the second moments and the velocity dispersion. Differentiating the second term on the left hand side, part of the result cancels part of the third term and we arrive *Jeans' equations* for a collisionless fluid,

$$\nu \frac{\partial \bar{v}_j}{\partial t} + \bar{v}_i \nu \frac{\partial \bar{v}_j}{\partial x_i} = -\nu \frac{\partial \Phi}{\partial x_j} - \frac{\partial}{\partial x_i}(\nu \sigma_{ij}^2) \quad (j=1,2,3). \tag{1.129}$$

$$\text{acceleration} + \begin{array}{c} \text{kinematic} \\ \text{viscosity} \end{array} = \text{gravity} + \text{pressure}$$

While in Cartesian coordinates all three Jeans equations have the same form, this is not true in other coordinate systems.

By setting the acceleration and shear terms to zero, we can recover the equation for hydrostatic equilibrium. Other uses include calculating the number density and potential self-consistently, assuming a given model for the velocity dispersion.

As an example, we will calculate the surface density of a sheet of fluid with plane parallel symmetry and reflection symmetry in the z-axis. The mass surface density, $\Sigma(z)$, is defined by

$$\Sigma(z) = \int_{-z}^{z} \rho(z')dz' \ . \tag{1.130}$$

Taking σ_{zz}^2 to be constant (iosthermal), the $j=3$ Jeans Equation gives

$$\frac{1}{\nu}\frac{\partial}{\partial z}(\nu \sigma_{zz}^2) = -\frac{\partial \Phi}{\partial z} = -2\pi G \Sigma(z), \tag{1.131}$$

where we have used Gauss' theorem to write the last part of the equality. The solution to (1.122) is of the form

$$\Sigma(z) = -\frac{\sigma_{zz}^2}{2\pi G}\left(\frac{d}{dz}\ln \nu\right). \tag{1.132}$$

If we consider the population of of nearby G and K dwarf stars, their measured velocity dispersion is found to be $\sigma_{zz}^2 = (20 \text{ km/s})^2$ and their scale height h, defined as

$$h^{-1} \equiv \frac{d}{dz}\ln \nu \tag{1.133}$$

is found to be approximately 300 pc. This gives a mass surface density of $\Sigma \approx 50\ M_\odot \text{pc}^{-2}$ for the disk of the Milky Way.

1.10. JEANS' EQUATION IN SPHERICAL COORDINATES

Applying the same equation to the Mestel disk model, we should be able to calculate a surface density from

$$v_c^2 = 2\pi G \Sigma_{\text{Mestel}} R_0, \qquad (1.134)$$

where measured values of v_c^2 and R_0 for the Milky Way are $(220 \text{ km/s})^2$ and 8 kpc, respectively. Substituting, we find that the ratio of the predicted Mestel disk surface density and that of the perpendicular plane model is

$$\frac{\Sigma_{\text{Mestel}}}{\Sigma_{\text{perp}}} = \frac{v_c^2}{\sigma_{zz}^2} \cdot \frac{h}{R_0} = \left(\frac{220}{20}\right)^2 \cdot \left(\frac{0.3}{8}\right) \approx 4.5. \qquad (1.135)$$

The assumptions going into the Jeans analysis are involve considerably less extrapolation than those going into the Mestel analysis. In particular we have assumed that *all* of the mass of the Milky Way lies in the disk. This might at first seem reasonable, since all but perhaps 15% of the light in the Milky Way lies in the disk. But if the disk were embedded in a more nearly spherical component, this would affect the circular velocity but not in the perpendicular velocity dispersion. The currently favored interpretation is that some large fraction of the mass interior to the Sun's orbit lies in an unseen, dark component. We will see that clusters of galaxies and the cosmic microwave background (CMB) provide additional evidence for such dark matter. A great many investigators have searched for Milky Way dark matter in various forms: ionized gas, atomic gas, molecular gas, compact stars and planet sized condensations. All have proved unsuccessful. Observations of the CMB in particular lead us to believe that the dark matter is non-baryonic – that it is not composed of protons, neutrons and electrons.

1.10 Jeans' equation in spherical coordinates

We start by writing down the collisionless Boltzmann equation (1.119) in spherical coordinates (r, θ, ϕ):

$$\frac{\partial f}{\partial t} + \dot{r}\frac{\partial f}{\partial r} + \dot{\theta}\frac{\partial f}{\partial \theta} + \dot{\phi}\frac{\partial f}{\partial \phi} + \dot{v}_r\frac{\partial f}{\partial v_r} + \dot{v}_\theta\frac{\partial f}{\partial v_\theta} + \dot{v}_\phi\frac{\partial f}{\partial v_\phi} = 0, \quad (1.136)$$

where the time derivatives of the coordinates may be expressed in terms the velocity components,

$$\begin{aligned} \dot{r} &= v_r, \\ \dot{\theta} &= \frac{v_\theta}{r} \quad \text{and} \\ \dot{\phi} &= \frac{v_\phi}{r \sin \theta} \end{aligned} \quad (1.137)$$

Lagrange's equations give the components of the acceleration,

$$\begin{aligned} \dot{v}_r &= \frac{v_\theta^2 + v_\phi^2}{r} - \frac{\partial \Phi}{\partial r}, \\ \dot{v}_\theta &= \frac{v_\phi^2 \cot \theta - v_r v_\theta}{r} - \frac{1}{r}\frac{\partial \Phi}{\partial \theta} \quad \text{and} \\ \dot{v}_\phi &= \frac{-v_\phi v_r - v_\phi v_\theta \cot \theta}{r} - \frac{1}{r \sin \theta}\frac{\partial \Phi}{\partial \phi}. \end{aligned} \quad (1.138)$$

(Continued on next page.)

One then substitutes these into the CBE (1.136), multiplies by the powers of v_r, v_θ and v_ϕ, and integrates over velocity space to obtain moment equations. These give one Jeans equation for each component of the potential gradient (force term). The radial Jeans equation is

$$\nu \frac{\partial \bar{v}_r}{\partial t} + \nu \left(\bar{v}_r \frac{\partial \bar{v}_r}{\partial r} + \frac{\bar{v}_\theta}{r} \frac{\partial \bar{v}_r}{\partial \theta} + \frac{\bar{v}_\phi}{r \sin \theta} \frac{\partial \bar{v}_r}{\partial \phi} \right) + \frac{\partial}{\partial r}(\nu \sigma_{rr}^2) + \frac{1}{r} \frac{\partial}{\partial \theta}(\nu \sigma_{r\theta}^2) + \frac{1}{r \sin \theta} \frac{\partial}{\partial \phi}(\nu \sigma_{r\phi}^2)$$
$$+ \frac{\nu}{r}[2\sigma_{rr}^2 - (\sigma_{\theta\theta}^2 + \sigma_{\phi\phi}^2 + \bar{v}_\theta^2 + \bar{v}_\phi^2) + \sigma_{r\theta}^2 \cot \theta] = -\nu \frac{\partial \Phi}{\partial r} \quad . \quad (1.139)$$

This may seem daunting (especially when one considers that the polar and azimuthal equations are similarly lengthy), but in practice one typically makes a number of simplifiying assumptions before invoking the radial Jeans equation:

a) Steady-state hydrodynamic equilibrium implies $\frac{\partial}{\partial t} = 0$ and $\bar{v}_r = 0$.

b) Spherical symmetry implies $\bar{v}_\theta = \bar{v}_\phi = 0$, $\sigma_{r\theta}^2 = \sigma_{r\phi}^2 = \sigma_{\theta\phi}^2 = 0$, and a single tangential velocity dispersion ("temperature") $\sigma_{t1}^2 = \sigma_{\theta\theta}^2 = \sigma_{\phi\phi}^2$.

With these assumptions the radial Jeans Equation becomes considerably more manageable,

$$\frac{1}{\nu} \frac{\partial}{\partial r}(\nu \sigma_{rr}^2) + 2\frac{(\sigma_{rr}^2 - \sigma_{t1}^2)}{r} = -\frac{\partial \Phi}{\partial r} = -\frac{GM(r)}{r^2}, \quad (1.140)$$

which reduces to spherical hydrostatic equilibrium when the velocity dispersion tensor is isotropic and $\sigma_{rr}^2 = \sigma_{t1}^2 = \sigma^2$,

$$\frac{1}{\nu} \frac{\partial \nu \sigma^2}{\partial r} = -\frac{\partial \Phi}{\partial r} = -\frac{GM_r}{r^2} \quad . \quad (1.141)$$

To measure the departure from this condition, we define the *anisotropy parameter* β as

$$\beta \equiv 1 - \frac{\sigma_{t1}^2}{\sigma_{rr}^2} \quad (1.142)$$

which can take values in the range $-\infty < \beta < 1$, with the extremes corresponding, respectively, to purely circular and purely radial orbits. The spherical Jeans equation (1.140) can be rewritten to give an expression for the mass $M(r)$ within a radius r

$$M(r) = -\frac{r\sigma_{rr}^2}{G} \left[\underbrace{\frac{d \ln \nu}{d \ln r}}_{\approx -3} + \underbrace{\frac{d \ln \sigma_{rr}^2}{d \ln r}}_{\approx -0.2} + \underbrace{2\beta(r)}_{?} \right]. \quad (1.143)$$

The difficulty with determining $\beta(r)$ from observations, is that one can only measure the line-of-sight velocity dispersion σ_{los}^2. It is not in general possible to make independent measurements of σ_{rr}^2 and σ_{t1}^2 (see Figure 1.22).

The above table gives "typical" values of the velocity dispersion and scale length for a giant elliptical galaxy and a very rich cluster of galaxies. The cluster of galaxies will have of order 100 bright galaxies (and a great many very much fainter galaxies). We can use these values and equation (1.143)

1.10. JEANS' EQUATION IN SPHERICAL COORDINATES

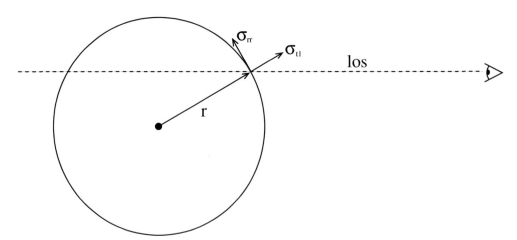

Figure 1.22: An observer measuring the Doppler broadening of lines in a galaxy spectrum sees some combination of the radial and tangential velocity dispersion, averaged over the line of sight.

system	σ_{rr}^2	N	R
galaxy	$(300 \text{ km/s})^2$	10^{11}	10 kpc
cluster	$(1000 \text{ km/s})^2$	10^2	1000 kpc

above to calculate a dynamical mass for a rich cluster. This can be compared to the dynamical masses computed for the constituent galaxies. The ratio of these two is

$$\frac{M_{\text{cluster}}}{N_{\text{gal}} M_{\text{gal}}} \approx \frac{1}{N_{\text{gal}}} \frac{\sigma_{\text{clus}}^2}{\sigma_{\text{gal}}^2} \frac{R_{\text{clus}}}{R_{\text{gal}}} \approx \frac{1}{100} \cdot 10 \cdot 100 = 10. \tag{1.144}$$

This very large discrepancy was first noted by Zwicky in the mid-thirties. It was at first known as the "missing mass" problem, but "missing light" would have been more correct, as the mass was surely present. For the next 40 years this problem was given scant attention, or dismissed as the result of some combination of bad data and bad modeling. When the first X-ray observations of clusters were made in the 1970s, very different observations and modeling led to the same conclusion. The missing mass problem became part of the larger "dark matter" problem that we first encountered within the Milky Way. Thirty years of effort have failed to produce a non-gravitational detection of this dark matter. In the meantime evidence (to be described later) has mounted that this matter must be non-baryonic.

Starting with the hydrostatic equation (1.141) and making the additional assumption of *isothermality*, taking the velocity dispersion $sigma^2$) to be independent of radius, allows us to move the velocity dispersion outside the derivative. Taking all particles to have the same mass m the density is then $\rho = m\nu$ and the hydrostatic equation reduces to

$$\frac{\sigma^2}{\rho} \frac{\partial \rho}{\partial r} = -\frac{\partial \Phi}{\partial r} = -\frac{GM_r}{r^2} \ . \tag{1.145}$$

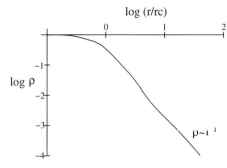

Figure 1.23: A schematic representation of the density profile of an isothermal sphere.

Multiplying this by r^2, and differentiating with respect to r and then multiply by $1/r^2$ we get

$$\frac{1}{r^2}\frac{d}{dr}\left(\frac{r^2}{\rho}\frac{d}{dr}\rho\right) = -\frac{4\pi G\rho}{\sigma^2} \quad . \tag{1.146}$$

Substituting $\rho \equiv \lambda e^\psi$ (with the central density given by $\rho_c = \lambda$), we get

$$\left(\frac{\sigma^2}{4\pi G\lambda}\right)\frac{1}{r^2}\frac{d}{dr}\left(r^2\frac{d}{dr}\psi\right) = -e^\psi. \tag{1.147}$$

Making a change of variables $\xi \equiv \frac{r}{\alpha}$ with $\alpha^2 = \left(\frac{\sigma^2}{4\pi G\lambda}\right)$, we get

$$\frac{1}{\xi^2}\frac{d}{d\xi}\left(\xi^2\frac{d}{d\xi}\psi\right) = -e^\psi, \tag{1.148}$$

which looks much like the Lane-Emden equation for polytropic stellar models, except the right-hand-side is an exponential instead of a power law. Indeed it is the limiting case of the Lane-Emden analysis for infinite polytropic index. The solution to this equation is called the *isothermal sphere*. We are interested in solutions with boundary conditions $\psi(0) = \psi'(0) = 0$. Figures 1.23 and 1.24 show the density and circular velocity of the isothermal sphere as a function of radius. Note that for large r, they approach the asymptotic values first mentioned in Sec. (1.5):

$$\frac{v_c^2}{\sigma^2} = -\frac{d\ln\rho}{d\ln r} \to 2. \tag{1.149}$$

Note also that wiggles persist at large radii, though with smaller and smaller amplitudes.

1.11 Jeans' equation applied; Jeans' theorem

Many if not most galaxies appear to exhibit cylindrical symmetry. Following the outline given above for obtaining Jeans' equations in spherical coordinates, one can obtain the cyclindrical versions. The radial Jeans' equation in cylindrical coordinates has the form

$$\frac{\partial}{\partial t}(\nu \overline{v_R}) + \frac{\partial}{\partial R}(\nu \overline{v_R^2}) + \frac{\partial}{\partial z}(\nu \overline{v_R v_z}) + \frac{1}{R}\frac{\partial}{\partial \phi}(\nu \overline{v_R v_\phi}) + \nu \left(\frac{\overline{v_R^2} - \overline{v_\phi^2}}{R} + \frac{d\Phi}{dR} \right) = 0. \qquad (1.150)$$

(Continued on next page.)

1.11. JEANS' EQUATION APPLIED; JEANS' THEOREM

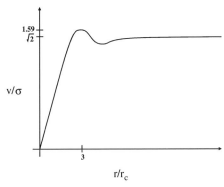

Figure 1.24: Circular velocity(normalized by the one dimensional velocity dispersion) as a function of radius in an isothermal sphere. The peak value occurs at $\frac{r}{r_c} \approx 3$

In the steady state the average azimuthal velocity $\overline{v_\phi}$ does *not* in general equal the circular velocity at that radius $v_c(R)$. The difference between the two is called the *asymmetric drift*,

$$v_a \equiv \overline{v_\phi} - v_c, \tag{1.151}$$

averaged over a volume element at a fixed position in space. Recall from the virial theorem [eqns. (1.38) and (1.40)] the relation

$$\langle v^2 \rangle_{\text{orbit}} = \langle v_c^2 \rangle_{\text{orbit}}, \tag{1.152}$$

where $\langle v^2 \rangle = \langle \overline{v_\phi^2} + \overline{v_R^2} + \overline{v_z^2} \rangle$. For a steady state $\overline{v_R} = \overline{v_z} = 0$. We then can substitute $\sigma_{RR}^2 = \overline{v_R^2}$ and $\sigma_{Rz}^2 = \overline{v_R v_z}$ into the radial Jeans equation (1.150) and find

$$2 v_a v_c = \sigma_{RR}^2 \left[\frac{\sigma_{\phi\phi}^2}{\sigma_{RR}^2} - 1 - \frac{\partial \ln(\nu \sigma_{RR}^2)}{\partial \ln R} - \frac{R}{\sigma_{RR}^2} \frac{\partial(\sigma_{Rz}^2)}{\partial z} \right]. \tag{1.153}$$

To first order, this gives an expression for the asymmetric drift at a fixed position,

$$\frac{v_a}{v_c} \sim \frac{\sigma_{RR}^2}{v_c^2}. \tag{1.154}$$

Observations of nearby stars in the Milky Way show a correlation between the amplitude of the asymmetric drift and the stellar type, which is in turn correlated with stellar ages. Older stars (G,K,M class) tend to have higher velocity dispersions, due to the fact that they have had a longer time to get "kicked around" by scattering with nearby stars.

The *second* moment of the CBE in cylindrical coordinates gives the relationship between the Oort constants and the epicyclic freqency cited without proof in Section (1.7)

$$\frac{\sigma_{\phi\phi}^2}{\sigma_{RR}^2} = \frac{\kappa_o^2}{4\Omega_o^2} = \frac{-B}{A-B}, \tag{1.155}$$

where A and B are the Oort constants defined in equations (1.114) and (1.115).

1.11.1 Jeans' theorem

Jeans' Theorem is so simple that one at first doubts whether it has any content:

- Any steady-state solution of the collisionless Boltzmann equation depends on the (6-dimensional) phase-space coordinates only through the (3) isolating integrals of motion in the galactic potential.

- Any function of the the integrals of motion yields a steady-state solution of the collisionless Boltzmann equation.

To prove the second part of Jeans' Theorem, consider an integral of the motion $I(\vec{x}, \vec{v})$, where the total time derivative of I is zero,

$$\frac{d}{dt}I[\vec{x}(t), \vec{v}(t)] = \vec{\nabla} I \cdot \frac{d\vec{x}}{dt} + \frac{\partial I}{\partial \vec{v}} \cdot \frac{d\vec{v}}{dt} = 0 \quad . \tag{1.156}$$

But since $d\vec{v}/dt = -\nabla\Phi$, then equation (1.156) is precisely the Collisionless Boltzmann equation (1.119) with I representing a solution without explicit time dependence. Now take a general function of the integrals of motion $g(I_1, ..., I_n)$. The time derivative of g is given by

$$\frac{d}{dt}g = \sum_{j=1}^{n} \frac{\partial g}{\partial I_j} \underbrace{\frac{dI_j}{dt}}_{=0} = 0, \tag{1.157}$$

so we find that g is *also* an integral of the motion and is therefore *also* a solution of the CBE.

The first part of Jeans' Theorem is evident from the fact that the distribution function $f(\vec{x}, \vec{v})$ is itself an integral of the motion. Also known as Liouville's Theorem, this is the same as saying that the phase space density is conserved along the orbit.

Here we present a few practical applications of Jeans' Theorem. For the case of spherical symmetry and an isotropic velocity dispersion, the density is a function only of radius r and the distribution function depends only on total energy:

$$\rho = \rho(r) \Leftrightarrow f = f(E). \tag{1.158}$$

For a spherical system with an anisotropic velocity dispersion tensor, the distribution function will depend on energy and total angular momentum:

$$\rho(r), \beta(r) \Leftrightarrow f(E, L^2). \tag{1.159}$$

If the potential is axisymmetric, the z-component of the angular momentum will be conserved along a given orbit, making L_z an isolating integral of motion. Distribution functions of the form $f(E, L_z)$ will give solutions to the CBE. Integrating over the distribution function one can calculate spatial densities $\rho(\vec{x})$ for spheroids or surface densities $\Sigma(\vec{x})$ for disks and (in both cases) mean azimuthal velocities $\overline{v_\phi}(\vec{x})$.

Up until this point we have been rather schematic and qualitative. How does one actually calculate the distribution function for an arbitrary potential? Consider the isotropic, spherically symmetric case where $f = f(E)$. Following Binney and Tremaine, we define the relative potential Ψ and relative energy ε as

$$\Psi = -\Phi + \Phi_o \tag{1.160}$$

1.11. JEANS' EQUATION APPLIED; JEANS' THEOREM

and
$$\varepsilon = -E + \Phi_o = \Psi - \frac{1}{2}mv^2. \tag{1.161}$$

Then the number density in space can be calculated from the distribution function by integrating over all of velocity space

$$\nu(\vec{x}) = \int_0^{v_{max}} f(\vec{x}, \vec{v}) d^3\vec{v} = 4\pi \int_0^{\sqrt{2\Psi}} v^2 f(\Psi - \frac{1}{2}v^2) dv, \tag{1.162}$$

where we have used the escape velocity $v_e = \sqrt{2\Psi}$ as the upper limit in the velocity space integral. Performing a change of variables $d\varepsilon = vdv$ gives

$$\nu = \int_0^{\Psi} f(\varepsilon) \sqrt{2(\Psi - \varepsilon)} d\varepsilon. \tag{1.163}$$

Differentiating with respect to Ψ gives the following

$$\frac{d\rho}{d\Psi} = 4\pi M \int_0^{\Psi} \frac{f(\varepsilon)}{\sqrt{2(\Psi - \varepsilon)}} d\varepsilon. \tag{1.164}$$

Applied mathematicians may recognize this as an *Abel integral equation*, which can be inverted to give the distribution function as a function of density,

$$f(\varepsilon) = \frac{1}{\pi^2 \sqrt{8}M} \frac{d}{d\varepsilon} \int_0^{\varepsilon} \frac{d\rho}{d\Psi} \frac{d\Psi}{\sqrt{\varepsilon - \Psi}}. \tag{1.165}$$

We might, for example, use the above method to compute the distribution function for an isothermal sphere. But instead we make an inspired guess for a distribution function,

$$f(\varepsilon) = \frac{\lambda}{\sigma \sqrt{2\pi}} \exp\left[-\frac{(\Psi - \frac{1}{2}v^2)}{\sigma^2}\right]. \tag{1.166}$$

Integrating this over velocity space we get

$$\rho = \lambda \exp\left(\frac{\Psi}{\sigma^2}\right), \tag{1.167}$$

and see where it takes us. Solving this for the relative potential Ψ and then invoking Poisson's equation we find

$$\nabla^2 \Psi = \frac{1}{r^2}\frac{d}{dr}r^2\frac{d}{dr}\Psi = -4\pi G\rho = \frac{\sigma^2}{r^2}\frac{d}{dr}\frac{r^2}{\rho}\frac{d\rho}{dr} = 4\pi\lambda \exp\left(\frac{\Psi}{\sigma^2}\right), \tag{1.168}$$

Letting $\alpha^2 = \left(\frac{\sigma^2}{4\pi G\lambda}\right)$ and $\xi = r/\alpha$ we recover the equation for the isothermal sphere, equation (1.148) – our inspired guess of $f(\varepsilon)$ has indeed given us the isothermal sphere.

1.12 Stability: Jeans mass and spiral structure

Until this point we have been concerned primarily with building equilibrium systems. We have not yet addressed the question of whether these equilibrium configurations are stable or unstable. The standard method for determining the stability of a system in equilibrium is to perturb it slightly. Either it will return to its previous configuration and oscillate around it (stable) or it will continue in the direction of the perturbation (unstable). The primary forces at work are gravity, which drives collapse, and pressure, which resists collapse. Since we have been treating the collection of stars as a collisionless fluid, it is not quite correct to think of a conventional fluid pressure, but for our purposes we take it that the stellar velocities produce a kind of effective pressure.

In the best of all possible worlds, one would begin with a time-independent distribution function and corresponding potential that solve the collisionless Boltzmann and Poisson equations. One then would consider small deviations of order $\epsilon \ll 1$ that can change as a function of time.

$$\begin{aligned} f(\vec{x}, \vec{v}, t) &= f_o(\vec{x}, \vec{v}) + \epsilon f_1(\vec{x}, \vec{v}, t) \\ \Phi(\vec{x}, t) &= \Phi_o(\vec{x}) + \epsilon \Phi_1(\vec{x}, t). \end{aligned} \tag{1.169}$$

Substituting (1.169) into the CBE and Poisson equation and equating terms of $\mathcal{O}(\epsilon)$, we get

$$\frac{\partial f_1}{\partial t} + \vec{v} \cdot \frac{\partial f_1}{\partial \vec{x}} - \vec{\nabla}\Phi_0 \cdot \frac{\partial f_1}{\partial \vec{v}} - \vec{\nabla}\Phi_1 \cdot \frac{\partial f_0}{\partial \vec{v}} = 0 \tag{1.170}$$

and

$$\nabla^2 \Phi_1(\vec{x}, t) = 4\pi G \int f_1(\vec{x}, \vec{v}, t) d^3\vec{v} \quad . \tag{1.171}$$

But as with our study of equilbria, solution for, $f_1(\vec{x}, \vec{v}, t)$, a function of 7 variables, is impractical.

We resort, instead, to perturbing Jeans' equations, and thus get the perturbations to low order moments of the distribution function, which ought to work reasonably well for large scale perturbations. But even Jeans' equation, the first velocity moment of the CBE,

$$\nu \frac{\partial \bar{v}_j}{\partial t} + \bar{v}_i \nu \frac{\partial \bar{v}_j}{\partial x_i} = -\nu \frac{\partial \Phi}{\partial x_j} - \frac{\partial}{\partial x_j}(\nu \sigma_{ij}^2).$$

presents difficulties. Six distinct components of the velocity dispersion tensor must be separately perturbed. Such a treatment is "beyond the scope of the present treatment." We therefore retreat to a fluid (as opposed to collisionless) treatment of the perturbations to Jeans' equations, assuming an isotropic velocity dispersion tensor. We multiply the number density ν by a particle mass, giving a density ρ. Dividing through by density and associating the $(\rho\sigma^2)$ term with pressure P gives the (collisional) fluid analog, the Euler equation:

$$\frac{\partial \vec{v}}{\partial t} + (\vec{v} \cdot \vec{\nabla})\vec{v} = -\vec{\nabla}\Phi - \frac{1}{\rho}\vec{\nabla}P. \tag{1.172}$$

The zeroth moment of the CBE gives the mass continuity equation

$$\frac{\partial \rho}{\partial t} + \vec{\nabla} \cdot (\rho \vec{v}) = 0. \tag{1.173}$$

1.12. STABILITY: JEANS MASS AND SPIRAL STRUCTURE

Just as with the distribution function above, we consider small deviations from the steady-state equilibrium solutions to the fluid equations (1.172) and (1.173). For small ϵ, we have the following expressions for the density, velocity, pressure, and potential of the fluid:

$$\begin{aligned}
\rho(\vec{x}, t) &= \rho_0(\vec{x}) + \epsilon \rho_1(\vec{x}, t) \\
\vec{v}(\vec{x}, t) &= \vec{v}_0(\vec{x}) + \epsilon \vec{v}_1(\vec{x}, t) \\
P(\vec{x}, t) &= P_0(\vec{x}) + \epsilon P_1(\vec{x}, t) \\
\Phi(\vec{x}, t) &= \Phi_0(\vec{x}) + \epsilon \Phi_1(\vec{x}, t)
\end{aligned} \quad (1.174)$$

The $\mathcal{O}(\epsilon)$ terms give

$$\frac{\partial \vec{v}_1}{\partial t} + (\vec{v}_0 \cdot \vec{\nabla})\vec{v}_1 + (\vec{v}_1 \cdot \vec{\nabla})\vec{v}_0 = -\vec{\nabla}\Phi_1 - \vec{\nabla}h_1, \quad (1.175)$$

$$\frac{\partial \rho_1}{\partial t} + \vec{\nabla} \cdot (\rho_0 \vec{v}_1) + \vec{\nabla} \cdot (\rho_1 \vec{v}_0) = 0, \quad (1.176)$$

and

$$\nabla^2 \Phi_1 = 4\pi G \rho_1, \quad (1.177)$$

where the term h_1 in equation (1.175) is the first-order perturbation to the *enthalpy*, defined as

$$h \equiv \int_0^\rho \frac{dP(\rho)}{\rho}, \quad (1.178)$$

giving the perturbation expression for the enthalpy as

$$h_1 = \left(\frac{dP}{d\rho}\right)_{\rho_0} \frac{\rho_1}{\rho_0} = v_s^2 \frac{\rho_1}{\rho_0}, \quad (1.179)$$

where v_s is the sound speed in the fluid.

To get Eqns. (1.175-1.177) into a form which can be solved analytically, Jeans made an additional simplifying assumption, that of a static, infinite background density ($\rho_0 = $ const, $\vec{v}_0 = 0$, $\Phi_0 = 0$). This assumption is *neither* physically realistic nor self-consistent, but it allows us to write the fluid equations as two coupled first-order partial differential equations:

$$\frac{\partial \rho_1}{\partial t} + \rho_0 \vec{\nabla} \cdot \vec{v}_1 = 0 \quad (1.180)$$

and

$$\frac{\partial \vec{v}_1}{\partial t} = -\vec{\nabla}h_1 - \nabla \Phi_1. \quad (1.181)$$

These can be combined into a single second-order equation in the density perturbation ρ_1

$$\frac{\partial^2 \rho_1}{\partial t^2} - v_s^2 \nabla^2 \rho_1 - 4\pi G \rho_0 \rho_1 = 0. \quad (1.182)$$

This is wave equation, so we try solutions of the form

$$\rho_1(\vec{x}, t) = C \exp[i(\vec{k} \cdot \vec{x} - \omega t)]. \quad (1.183)$$

Substituting (1.183) into (1.182) gives a dispersion relation

$$\omega^2 = v_s^2 k^2 - 4\pi G \rho_0. \tag{1.184}$$

If $\omega^2 > 0$, the perturbations propagate like sound waves through the fluid, i.e. the perturbations are *stable*. If $\omega^2 < 0$, we get a real, positive exponent in the equation (1.183), giving unstable exponential growth. Setting $\omega^2 = 0$ gives the critical wavenumber, or equivalently the critical length scale ($\lambda = 2\pi/k$), for which initial density perturbations will grow exponentially. This wavelength called the *Jeans length* and is given by

$$\lambda_J^2 = \frac{\pi v_s^2}{G \rho_0}. \tag{1.185}$$

The amount of matter within such a critical volume is easily calculated assuming a spherical density perturbation of radius λ_J and average density ρ_0. This *Jeans mass* is typical of the structure we can expect to be formed due to unstable gravitational collapse:

$$M_J = \frac{4\pi}{3} \rho_0 \lambda_J^3. \tag{1.186}$$

Our derivation thus far has been for *collisional* fluids. How are these results used to understand collisionless stellar systems? The speed of sound v_s^2 must be replaced by the velocity dispersion σ. This gives a dispersion relation that is slightly different from equation (1.184) but the Jeans mass remains unchanged.

For an isolated spherical system of point masses, the crossing time r/σ is approximately the same as the free-fall collapse time, $t \sim 1/\sqrt{G\rho}$. Nothing smaller than the whole system can be unstable to gravitational collapse.

Going back to the linear analysis, we can use the same methods with modifications appropriate to other geometries. Cylindrical coordinates would seem a good choice for fluid disks. We assume perturbations of the form $e^{im\phi}$ for integer values of m. The simplest case is that of $m = 0$, which gives the dispersion relation

$$\omega^2 = v_s^2 k^2 - 2\pi G \Sigma |k| + \kappa^2, \tag{1.187}$$

where Σ is the integrated surface density of the disk and κ is the same epicyclic frequency introduced in section (1.7). A dimensionless factor, *Toomre's Q*, defined by

$$Q \equiv \frac{v_s \kappa}{\pi G \Sigma} \tag{1.188}$$

allows the dispersion relation to be written as

$$\omega^2 = Q^2 k^2 \left(\frac{\pi G \Sigma}{\kappa}\right)^2 - 2\pi G \Sigma |k| + \kappa^2. \tag{1.189}$$

In this form the stability of the system depends solely on Q:

$$\begin{aligned} Q &= 1 & &\text{gives perfect square, } \omega^2 \geq 0 \\ Q &> 1 & &\text{stable for all } k \\ Q &< 1 & &\text{unstable for some } k \end{aligned} \tag{1.190}$$

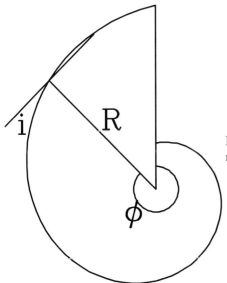

Figure 1.25: The crest of an $m = 1$ perturbation makes angle i with the tangential direction.

A similar treatment of a collisionless stellar disk yields a similar dispersion relation, with the qualification that the dimensionless constant π used in defining Toomre's Q, (eqn. 1.188), is replaced by the dimensionless constant 3.36, a change of less than 10%.

For values of $m = 1, 2, ...$, the picture gets somewhat more complicated. The first complication is that the notation changes – we switch from ω to $\Omega_p \equiv \omega/m$ where the subscript p denotes pattern speed. Full derivation of the dispersion relation is again "beyond the scope of the present treatment." There are nonetheless some important points that are relatively easy to follow.

First, there is a problem. Suppose that the stars in a disk are in circular orbits, with some orbits overpopulated so as to cause an apparent spiral. The position of the "crest" in polar coordinates is described by

$$\phi(R, t) = \phi_0 + \Omega(R) t \quad . \tag{1.191}$$

The the crest will in general make an angle i with the tangential direction. As illustrated in Figure 1.25 we find

$$\cot i = \left| R \frac{\partial \phi}{\partial R} \right| = R \left| \frac{d\Omega}{dR} \right| t \quad \text{whence} \tag{1.192}$$

$$\cot i = \left| \frac{d\ln\Omega}{d\ln R} \right| \Omega t \quad . \tag{1.193}$$

The relevant time t is of order the Hubble time, in which case the product Ωt is of order 100π. If $\Omega(R)$ is not constant (solid body rotation) any spiral that is more than a few galactic "years" old will have a very small inclination i – it will have wound up.

Lindblad came up with a scheme that partially mitigates the wind-up problem. Recall that with a few special exceptions, the orbits in a circularly symmetric potential do not close – the radial and aziumthal periods, T_r and T_a are different. There is a phase $\Delta \phi \equiv \omega_a T_r$ which is not in general a multiple of 2π, nor is it in general an integer fraction thereof. But in a frame of reference rotating

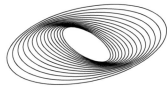

Figure 1.26: A series of nested ellipses produce a spiral arm. Such a "kinematic" spiral can only persist if there is a rotating frame in which all of the orbits are closed ellipses. This is *not* possible for a logarithmic potential.

Figure 1.27: A star executes an elliptical epicycle as a spiral arm winds up. The coordinate system is in the frame of guiding center, which is moving toward the left. Notice that the star spends a large part of the epicycle close to the spiral arm.

with orbital frequency Ω_p, the phase of the orbit is given by

$$\Delta\phi_p = \Delta\phi - \omega_p T_r \quad . \tag{1.194}$$

If $\Delta\phi_p = 2\pi n/m$ the orbit closes afer m radial oscillations and

$$\Omega_p = \frac{\Delta\phi}{T_r} - \frac{\Delta\phi_p}{T_r} = \omega_a - \frac{n}{m}\omega_r \approx \Omega - \frac{n}{m}\kappa \quad , \tag{1.195}$$

where κ is the epicyclic frequency. Taking $n = 1$ and $m = 2$ gives closed ellipses. Let us suppose that there is a range of radii for which $\Omega - \kappa/2$ is constant. If one could then somehow arrange to populate at set of perturbed orbits with the correct phase, one might produce a pattern like that in Figure 1.26 that would persist for some time.

Lindblad's construction may point in the right direction, producing roughly the correct pattern speed, but it does not work well for our flat rotation curve Mestel disk model, for which $\kappa = \sqrt{2}\Omega$ giving $\Omega_p = 0.29\Omega$. Spirals wind up more slowly, but they still wind up.

But Lindblad's construction is kinematic, in that the spiral is an accident of initial condition rather than a dynamical entity. Various investigators have attempted to construct persistent spiral patterns for which self-gravity in some way overcomes the wind-up problem. Chief among these have been Lin and his collaborators, who have advanced the scenario of quasi-stationary *density waves*. For many galaxies the wind-up problem appears to preclude quasi-stationary density waves, but they may be possible in galaxies in which a bar forces them.

An alternative approach is to give up on the quest for immortality to embrace short-lived spirals. Toomre in particular has emphasized that close encounters between galaxies will cause tides that produce magnificent transient spiral patterns. The *swing amplification* of leading spiral arms illustrated in Figure 1.27, plays a major role in this model. Measuring the inclination i of the spiral arm with respect to the tangential (horizontal) direction we have

$$\cot i = -R\frac{d\Omega}{dR}t \quad , \tag{1.196}$$

where $\frac{d\Omega}{dR} < 0$. Differentiation with respect to time and using the definitions of Oort's constants

1.12. STABILITY: JEANS MASS AND SPIRAL STRUCTURE

Equations 1.114 and 1.115, we that the time rate of change of the inclination angle is given by

$$\frac{di}{dt} = \frac{2A}{1 + A^2 t^2} \quad . \tag{1.197}$$

For a Mestel disk the epicyclic frequency is $\kappa = \sqrt{2}A$, roughly equally to the time rate of change of the inclination for most of an epicycle. Stars whose motions are in phase with the spiral arm stay in phase and self-gravitate, increasing the amplitude of the disturbance. Eventually the arm winds up.

Tides play a role, in particular, in the "grand design" spirals, starting with M51 for which the companion responsible for the tide is clearly visible, and M81, whose spiral arms appear to result from an encounter with M82, which itself is undergoing a starburst.

Yet another scheme for producing spiral arms is *self-propagating star formation*. The idea is that star formation propagates along an interface much like a detonation front, which is sheared out by differential rotation. This produces patterns more like those seen in the "flocculent" spirals. The bottom line on spiral structure is that different mechanisms may be at work in barred, grand design and flocculent spirals.

Impressum
Copyright: © 2013 Dr. Patrick Petter
Druck und Verlag: epubli GmbH, Berlin, www.epubli.de
ISBN 978-3-8442-7198-0